Islands for Offshore Nuclear Power Stations

Islands for Offshore Nuclear Power Stations

Binnie & Partners
London

A report prepared for the Commission of the European Communities, Directorate-General for Science, Research and Development.

Graham & Trotman
for the Commission of the European Communities

Published in 1982 by
Graham & Trotman Limited
Sterling House, 66 Wilton Road
London SW1V 1DE, United Kingdom

Reprint of the original edition 1982

for the Commission of the European Communities, Directorate-General
Information Market and Innovation

EUR 7534

© ECSC, EEC, EAEC, Brussels and Luxembourg, 1982

ISBN-13: 978-0-86010-373-8   e-ISBN-13: 978-94-009-7369-5

DOI: 10.1007/978-94-009-7369-5

Legal Notice

Neither the Commission of the European Communities, its contractors nor any person acting on their behalf; make any warranty or representation, express or implied, with respect to the accuracy, completeness, or usefulness of the information contained in this document, or that the use of any information, apparatus, method or process disclosed in this document may not infringe privately owned rights; or assume any liability with respect to the use of, or for damages resulting from the use of any information, apparatus, method or process disclosed in this document.

All rights reserved. No part of this publication may be reproduced, stored in a retrieval system, or transmitted in any form or by any means, electronic, mechanical, photocopying, recording or otherwise, without the prior permission of the publishers.

CONTENTS

		Page No.
	FOREWORD	(ix)
1.	**TERMS OF REFERENCE**	1
2.	**INTRODUCTION**	3
3.	**LITERATURE REVIEW**	3
4.	**DESIGN CRITERIA**	5
	Introduction	5
	Siting categories	5
	Design life	7
	Hazard	9
	Design parameters	10
	Nuclear power station	14
	Access	15
	Construction & operation personnel	16
	Transmission links	17
5.	**ISLAND CONCEPTS**	19
	Introduction	19
	Types of island	19
	Features of island types	21
	Comparison of island types	28
	Comparison of concrete & steel	30
	Conclusions	31
6.	**OUTLINE DESIGNS**	33
	Introduction	33
	Floating island	33
	Semisubmersible	37
	Gravity platform	41
	Caisson island	43
	Fill island	45
	Caisson retained island	55
	Protected floating island	57
	Caisson/fill island	59
	Caisson/piled island	59
	Cooling water	61
	Access	61

                                                                    Page No.

7. **CONSTRUCTION METHODS AND TIMES**   65

    Introduction   65
    Floating island and semisubmersible   65
    Gravity platform   73
    Caisson island   73
    Fill island   75
    Caisson retained   76
    Protected floating island   77
    Caisson/fill and caisson/piled islands   77
    Power station development   77

8. **PLANT CONSTRUCTION COSTS**   79

    General   79
    Floating island   79
    Semisubmersible   79
    Fixed island   80
    Fill island   80
    Caisson & composite islands   80
    Summary   80

9. **UNIT COSTS FOR ISLANDS**   81

    General   81
    Concrete in islands   81
    Sandfill   82
    Natural gravel   82
    Rockfill   83
    Filter   83
    Rock armour   84
    Concrete in armour units   84
    Sheet piled quays   84
    Summary   84

10. **COST ESTIMATES**   87

    General   87
    Costs of outline designs   87
    Auxiliary structures   95
    Access   97
    Transmission links   100
    Summary   101

		Page No.
11.	CONCLUSIONS	107
	Design criteria	107
	Types of island	107
	Siting categories - island suitability	109
	Timescale	111
	Production/construction facilities	111
	Factors affecting overall costs	111
	Power station weight and layout	111
	Summary	111
12.	RECOMMENDATIONS	113

APPENDICES

A	TECHNICAL ANNEX TO CONTRACT NO 623-79-12 EC1 UK
B	DRAFT INFORMATION NOTE
C	SELECTED BIBLIOGRAPHY
D	WATER DEPTHS WITHIN STUDY AREA
E	WAVE HEIGHTS WITHIN STUDY AREA
F	WAVE FORCES
G	TIDAL RANGES WITHIN STUDY AREA
H	EXTREME WATER LEVELS
J	DYNAMIC STABILITY OF OFFSHORE ISLANDS
K	COOLING WATER DISPERSION

FOREWORD

The Working Group "Offshore Nuclear Power Plants" was established by the Commission of the European Communities in 1971. At that time it appeared that the option offered by creating new sites offshore should be explored. The present development of nuclear power has not, however, reached the stage where offshore siting can be considered as a near or medium term proposition.

The purpose of this report is to summarize the state of the art for the provision of floating or fixed structures, or man-made islands of the size needed for the construction of nuclear and other power stations. It describes the main factors which must be taken into account in the design and location of such islands and provides an indication of feasibility and cost for each design at the present time. It deals mainly with the civil engineering problems and not other major factors, such as the law of the sea, the rights of countries to locate nuclear establishments off their coast, their safety, security, the energy connection with the mainland, the marine ecology, the logistics, etc.; many of these problems are directly related to the site and have major economic and political implications. By studying application examples the Working Group has occupied itself with these other factors.

The present report, prepared by Binnie and Partners under contract to the Commission of the European Communities and with the technical support of the Offshore Group, summarizes the major technical challenges to the conttstruction of offshore locations for power stations and it is hoped it will form part of a basis on which to carry out the site specific research necessary if offshore power stations are required in the Community.

The Offshore Group wishes to thank Binnie & Partners for their good cooperation and procedure in preparing the report, and the services of the Commission's secretariat, without whose help the work could not have been concluded.

1. TERMS OF REFERENCE

1.1   Our Terms of Reference are contained in the technical annex to our Contract No. 623-79-12EC1 UK with the European Atomic Energy Community represented by the Commission of the European Communities. A copy is attached as Appendix A.

1.2   The scope of the study as set out in the technical annex is as follows:

- to review alternative methods of island construction and their application to nuclear power station siting;

- to relate alternative methods of construction to three basic categories of site and to report on the alternatives that are available for each category;

- to provide a broad based economic comparison of the alternatives in each category for different plant capacities;

- to provide the timescale for construction of each alternative;

- to highlight those factors such as material availability, production and construction facilities that could influence the preferred alternative in each category and the economic implications.

1.3   The three different categories of site defined in the technical annex are discussed in Section 4.3.

1.4   The technical annex provided for the Working Group on Offshore Nuclear Power Plants to supply further data for use in the study. At the 13th meeting of the Working Group on the 10th January 1980 a draft information note was given to us. This is attached as Appendix B. The most importants points made in the note are that:

- because gas cooled reactors are larger and heavier than water cooled reactors they are unsuitable for floating or fixed platform type islands. For the study, the data applicable to a water cooled reactor should be used.

- A basic development of 2500 MW in units of 1250 MW is to be considered but with the possibility of extension to 5000 MW and 10000 MW.

- The main plant for each 1250 MW unit can be accommodated within a rectangular area of 250 x 200 m. The auxilliary structures for a land-based power station require a further 10 ha, giving a total of 15 ha. This area could be reduced for offshore power stations.

- The study should assume an area of 5 ha for the main plant for a 1250 MW unit and should consider the implications of a range of platform sizes in multiples of 5 ha up to 15 ha.

Other design parameters for the plant were included in the information note.

1.5   During the discussion at the meeting on the 10th January it was agreed that:

- the area for the main plant buildings should be modified to 6 ha;
- the study should not be site specific.

1.6   Following the comments of the CEGB on the Second Interim Report (May 1980) we considered it would be useful to study, briefly, the effect on island design of nuclear plants of smaller size and lower weight than specified in the information note. This work is summarised in paragraphs 10.28–10.35.

2. INTRODUCTION

2.1   An offshore island for a nuclear power station must have an exceptionally high degree of security under all circumstances. Both the design criteria to be adopted and the design solutions necessary to meet them must be exhaustively examined. The examination of specific sites is not in the brief for this study. We have therefore attempted to make the study as broad as possible in order to cover a wide range of possible sites within three basic categories. The design criteria, including the design parameters, are discussed in Section 4.

2.2   The recent expansion of the offshore oil industry has initiated considerable development in the application of existing technology and research into new technology. In other areas the pressure on land resources has stimulated interest in land reclamation and artificial islands as an alternative to conventional industrial development. Thus many types of island have been considered for various special applications in the above areas of interest. These types of island and their suitability for nuclear power stations are reviewed in Section 5 - Island concepts - which identifies the most suitable types for the three siting categories.

2.3   Outline designs are developed in Section 6 for these types of island and construction methods and time scales in Section 7. The relative merits and feasibilities of the islands are assessed in these sections. Where existing technology is insufficient to guarantee the feasibility of an island, this is noted.

2.4   The basis of the costing of the islands is considered in Section 9 and the effects on power plant construction costs in Section 8. A broad based economic comparison of the islands is given in Section 10 and our conclusions and recommendations in Sections 11 and 12.

2.5   The effects on island costing of nuclear power plants of smaller size and lower weight is considered in Section 10 (10.28–10.35). This comparison, although outside the Terms of Reference, is included to indicate the possible savings in the cost of the islands if radical redesign of the nuclear power plant were envisaged.

2.6   The terminology used in the report for describing the various parts of the developments are given at the end of Appendix B.

2.7   This final report includes the findings of the two interim reports and is intended to be read without reference to them.

3. LITERATURE REVIEW

3.1   A review of the published literature on offshore island technology was carried out to supplement "in-house" knowledge on the construction of all types of islands. A selected bibliography is attached as Appendix C. Articles on planned and constructed islands in Europe, Canada, the United States of America and Japan are included. References are also given to works on design criteria and parameters where these are relevant.

3.2   The literature review has been supplemented by visits to discuss studies undertaken in Holland and Belgium. Further information and comments on the two interim reports have been received from the United Kingdom, France, Belgium and Germany. A schedule of correspondence is attached as Appendix D.

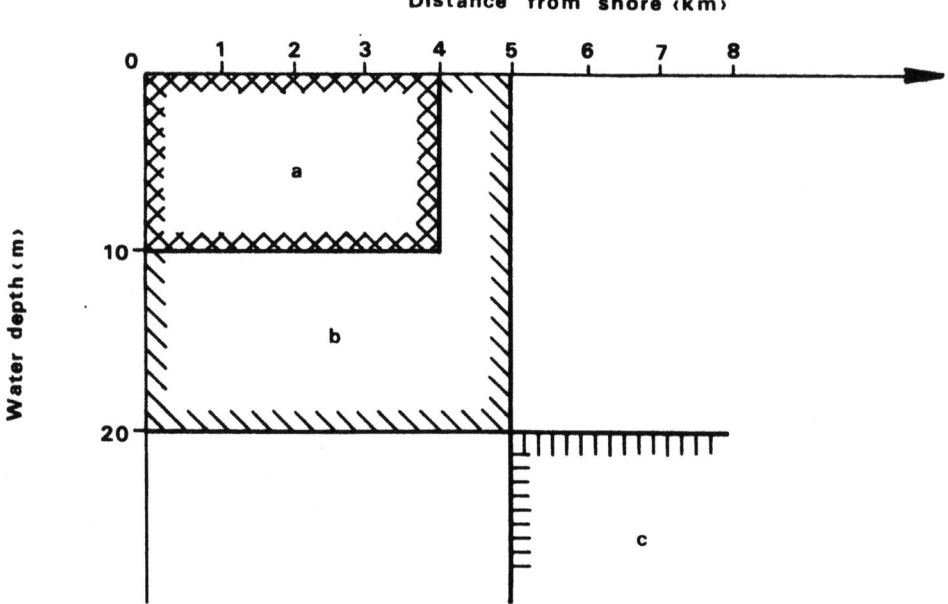

Siting categories a, b, c suggested in technical annex

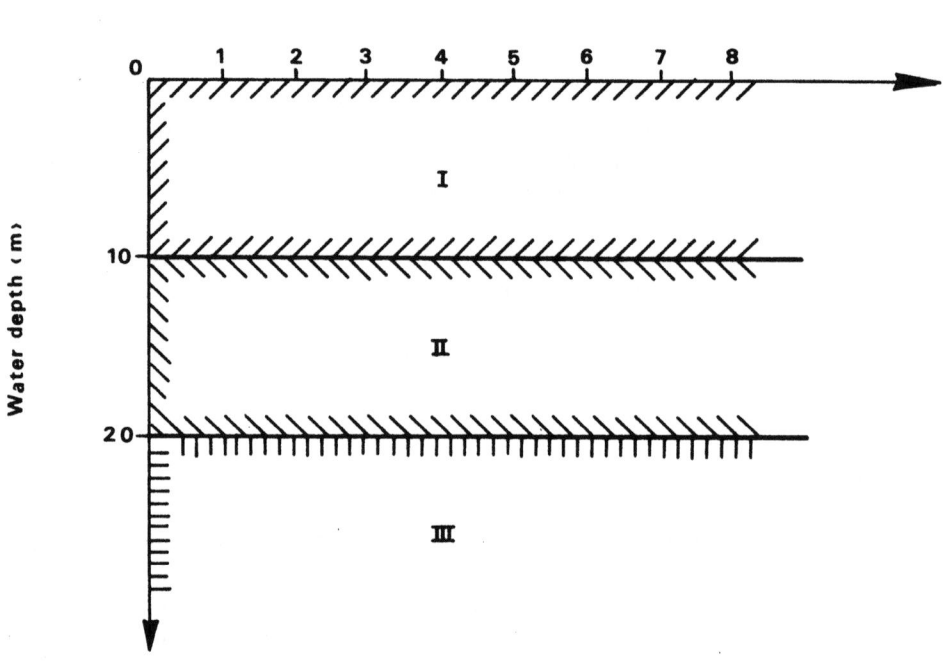

Siting categories I, II, III used in study

Siting categories   Drawing 1

4. DESIGN CRITERIA

Introduction

4.1   We have tried to make the study as broad as possible within our terms of reference. A full range of sites from tidal zones to deep water sites at the edge of the continental shelf have been considered. Although the shallower, more sheltered sites are more attractive than the deeper, more exposed ones, we have tried to relate existing technology to all these sites to determine the feasibility of their use and types of island that might be suitable.

4.2   The classification of the sites and some physical and legal limitations are discussed in Sections 4.3 - 4.15. The required design life for an island is discussed in Sections 4.16 - 4.33 and the problems of natural and man-made hazards in Sections 4.34 - 4.36. The range of design parameters that must be taken to cover the various sites is examined in Sections 4.37 - 4.62 and power station layout and weights, personnel and transmission links in Sections 4.63 onwards.

Siting categories

General

4.3   Our terms of reference state that the study shall not be site specific. Instead, three basic "siting categories" are suggested (see Appendix A). The relationship of water depth and distance from the shore for these suggested categories is shown in Drawing 1. It can be seen that deep sites close to the shore and shallow sites far from the shore are excluded.

4.4   As the water depth is the most important parameter in defining the type of island we prefer to designate siting categories in terms of water depth alone. The siting categories used in this report are:

Category	Water depth
I	less than 10 m
II	10 m - 20 m
III	greater than 20 m

4.6   These siting categories I, II and III cover all possible sites, (see Drawing 1).

4.7   An indication of the availability of the siting categories at various distances offshore is given in Appendix E, Drawings E1, E2 and E3.

4.8   The range of design parameters that can be expected at these sites is given in Table 1.

4.9   The conditions that may be met in the categories suggested both in the Technical Annex and in the revised siting categories adopted for use in this study will vary widely. Specific sites will have problems and combinations of problems that may be unique and cannot be fully explored within the framework of this study. We have therefore had to assume a broad range of design parameters for each category to ensure that the most likely conditions are covered.

4.10   Investigation of specific sites will require detailed data collection and analysis, model tests, design etc. which are beyond the scope of this study.

Physical limitations

4.11   The availability of the sea bed for the construction of islands is limited by its existing uses. The following factors are among those that may exclude possible sites:

> shipping lanes
> oil and gas fields
> oil and gas pipelines
> telephone cables
> military areas.

4.12   In addition to the above constraints the effect of the island on the ecology and environment will be a major consideration for inshore sites. For offshore sites these problems will be reduced although fish spawning grounds would have to be considered.

Legal aspects

4.13   At present there is no internationally accepted body of law defining the jurisdiction of coastal nations over their adjacent sea and sea beds. The United Nations Law of the Sea Conference has not yet obtained general acceptance for its proposed Convention of the Law of the Sea. However, the proposals are comprehensive and their implications must be considered in relation to possible sites for offshore islands.

4.14   The proposals split the sea and sea bed adjacent to a coastal nation into five zones:-

Zone	Distance Offshore (km)
Territorial Sea	0 - 12
Contiguous Zone	12 - 36
Exclusive Economic Zone	12 - 200 (max)
Continental Shelf	varies
Continental Margin	varies

4.15   According to the proposals a coastal state may establish and use artificial islands and other structures anywhere within its territorial sea and exclusive economic zone including the contiguous zone. Other states have freedom of navigation and the right to lay submarine cables and pipelines in the coastal state's exclusive economic zone but not in its territorial sea.

Design life

General

4.16   An island has to be designed to support and protect its power station(s) to a very high degree of security during its operating life. In addition the island must be capable of maintaining security until such time as the reactors are dismantled and removed or an alternative way is found of providing the required security.

4.17   The length of the design life assumed for islands must take into account the power station's:

>construction period
>operational life
>subsequent use/demolition.

Construction period

4.18   For a single development of 2,500 MW the construction period before the first unit becomes operational could be of the order of ten years. For a large development of 10,000 MW the construction could extend over fifteen or more years. The last unit might not become operational until the first unit had been running for ten years.

Operational life

4.19   The operational life of a PWR nuclear plant has not yet been fully established - a range from 25 to 50 years seems possible. We have taken an operational life of 40 years for the purpose of this study.

Subsequent use/demolition

4.20   At the end of its operational life the plant can be treated in several ways:

(a)   immediate demolition

(b)   isolation of the reactor followed by demolition within 30 years

(c)   isolation of the reactor and confinement of its parts followed by demolition within 100 years.

4.21   After demolition the radioactive materials can either be permanently encapsulated at the site or removed elsewhere.

4.22   The cost of immediate demolition is estimated to be some 2 - 2.5 times the cost of the other options. However, demolition is eventually required because the isolation and confinement are only safe for the period noted. Demolition after 30 years may produce the most economic solution.

4.23   The cost of demolition of the nuclear plant can be valued against the cost, if feasible, of maintaining the island for radioactive storage. Environmental aspects will also have to be considered.

Range of design life

4.24   It can be seen that the required design life could vary considerably depending on the chosen method of demolition. If, in addition, it is required to store the radioactive materials on the island then the design life may be required to be extremely long.

4.25   The possible range of design lives could be:-

	minimum	average	maximum
construction	5	10	15
operation	25	40	50
isolation	-	25	100
demolition	5	5	5
	35 years	80 years	165+ years

4.27   There is insufficient experience to be confident that design lives in excess of 80 years are feasible at present. Complete rebuilding of the exterior of the island would have to be allowed for if such design lives were required although in the event this rebuilding might not be required. However, only those types of islands which can be rebuilt in this way should be used for design lives of more than 80 years.

4.28   If materials are carefully chosen we consider that a design life of 50 years is feasible for all types of island.

Lengthening the design life

4.29   When designing for a longer life, problems of long term change are encountered. These can be split into:

> deterioration of the structure
> changes in the local environment.

4.30 Deterioration of the structure would be caused by factors such as corrosion of steel elements, degradation of concrete and breakdown of protective coverings. Long term experience of structures in the North Sea oil fields is only now becoming available and corrosion problems have proved more severe than expected.

4.31 Major maintenance work on the external structure of some types of island may be impossible unless the islands can be moved to specially prepared docks. For safety reasons this may not be possible. Unless such islands are capable of surviving their design lives without major maintenance their feasibility may be doubtful. Thus the effect of lengthening the required design life will be to reduce the choice of types of island. The maintenance costs of those that do remain feasible will, of course, be increased.

4.32 The local environment of the island will change with time. These changes, which may or may not be caused by the presence of the island, could be important at some sites. Examples of such changes are erosion or accretion, other man-made developments and changing standards of environmental acceptability. Although such changes are not likely to affect the feasibility of extending the design life of an island, they would have to be considered in more detail.

Summary

4.33 We have taken a design life of 80 years although some types of island cannot at present meet this requirement. For such islands either a reduced operational life or demolition immediately after final closedown may be required. The reason for this constraint is the lack of the experience required to predict with confidence that unexpected deficiencies will not be discovered. Longer experience of North Sea and similar structures and more research may allow this constraint to be removed.

Hazard

4.34 An island supporting a nuclear power station must be designed to have a high degree of safety even for exceptional events. For natural phenomena such as severe storms, earthquakes etc., this degree of safety can be achieved by using probability theory to extrapolate existing data to determine the magnitude of extreme events. Corporate or even national decisions may be needed to decide the risk of damage that is acceptable during the life of a power station.

4.35 For man-made hazards such as ship collision, aircraft impact, sabotage or military action, the same corporate or national decision will probably be required. For inshore locations, islands with greater security against these hazards are available. However, as the water depth increases, it becomes difficult to provide the same degree of protection and the cost of doing so increases rapidly.

4.36 The susceptibility of islands to hazards of this kind is discussed more fully in the sections dealing with the individual types of island.

Design parameters

General

4.37   The design parameters for use in the study are set out and discussed below. The range of parameters being considered for the three siting categories specified in Section 4.4 are listed in Table 1.

Table 1 Site conditions

	Category I	Category II	Category III
Water depth (m)	10 (maximum)	10 - 20	20 (minimum)
Distance from high water mark (km)	unlimited	unlimited	unlimited
Design wave height H (m)	3 - 6	3 - 10	6 - 15
Design wind, 3 second gust (m/s)	40	45	50
Tidal range (m)	0 - 9	0 - 9	0 - 9
Currents (m/s)	0 - 2	0 - 1.5	0 - 1

Water depth

4.38   The maximum water depth considered has not been limited. For the purpose of this study water depths are measured from mean sea level.

4.39   Many of the parameters become more severe with increasing water depth and design problems and construction costs rise disproportionately. Recent technology has enabled structures to be founded in great depths but the need for such structures has been determined by the presence of oil and gas in these deeper waters. An authority planning a location for an offshore nuclear power station is not constrained in the same manner. It appears unlikely that a site in water deeper than, say, 40 m would be required for environmental or other reasons. However, we have considered greater depths.

Distance from high water mark

4.40   The siting categories we have adopted are not limited to particular distances from high water mark. However, category I sites are likely to be close to the shore. Category II and III sites will normally be at increasing distances offshore. Transmission links from islands to the shore are expensive and locations far offshore bear a heavy financial penalty. Access for personnel to remote locations becomes a severe problem. These factors are discussed in Section 4.67.

Waves

4.41   Two concepts are important when considering the effects of waves. The first is the cumulative destructive power of the waves. The second is the maximum wave that will occur. The cumulative destructive power of an irregular series of waves can be assessed in terms of the "significant wave height" (H_s) of the series. The significant wave height is the mean height of the highest third of the waves in the series. The "maximum wave height" (H_{max}) of a series of waves is a multiple of the significant wave height. The ratio of H_{max} to H_s depends on the number of waves being considered but is generally about two.

4.42   An island must be designed to resist, without much damage, both the cumulative destructive power and the effects of the highest waves in the worst storm that may occur during its life. The magnitude of this worst or "design storm" will depend on the particular site and the return period chosen for the storm.

4.43   For the purposes of this study we have defined a "design wave height" (H_D) as the significant wave height of the design storm. The duration of the storm has been taken as 10,000 waves, equivalent to about 30 hours. For this duration of design storm the maximum wave height (H_{max}) will be about 2.1 times as high as the design wave height. This relationship has been used when fixing levels for the islands.

4.44   The water depths in siting categories I and II will limit the height of wave that can remain unbroken. In the deep water of category III depth will not limit the height of waves.

4.45   Appendix F gives values of design wave height (H_D) for particular offshore areas and notes the effects which modify waves as they progress shorewards.

4.46   Wave loadings generally provide the largest horizontal forces on the structures. Accurate methods of computing wave forces are complex and costly, often requiring the use of large computer programs. For our outline designs, simplified methods are sufficient to enable preliminary cost comparisons to be made between the types of island. More precise methods would be necessary only for a detailed, site specific study. Details of the methods used are given in Appendix G.

4.47   The stability of islands will be affected by both the height and the period of the waves. The relationship between the wave period and the natural frequency of the island is critical in determining the response of the island and indeed its feasibility because excessive movement is unacceptable. The dynamic behaviour of islands is discussed further in Section 6 and Appendix K.

Winds

4.48 For floating and fixed islands, wind loads affect the design of the island as well as the design of the plant it supports. In calculating forces due to wind the methods given in the British Standards Institute Code of Practice CP3 have been used.

Tidal range

4.49 The tidal range affects not only the mooring of floating islands and the extent of their freeboard but also the wave protection on non-floating islands. The tidal range varies considerably around the coast of Europe. In the Severn estuary and the Golfe de St. Malo it is as much as 11 m. Along the shores of the English Channel (La Manche) and the east coast of the United Kingdom it varies from 3 m to 8 m. Elsewhere in the North Sea it is less than 3 m and in the Mediterranean it is less than 1 m. An indication of the geographical variation of tidal range is given in Appendix H, drawings H1, H2 and H3.

Extreme water levels

4.50 Extreme water levels can result from abnormal weather conditions (storm surges) or earthquakes (tsunamis). These are considered in Appendix J.

4.51 The extreme water level is a site specific parameter and in order to reduce the number of variables a constant allowance of 2 m above the highest tide level has been assumed for design and costing purposes.

Currents

4.52 Currents are mainly caused by tidal variations and to a lesser degree by winds. The main problem caused by currents is scour of the seabed. They also will affect towing of any large floating structures and to a lesser extent their moorings. Conversely the lack of currents creates problems with the dispersal of the cooling water. This is discussed in more detail in Section 6.12 and Appendix L.

Earthquake and other disturbances

4.53 The design earthquake loading for the study has been specified (see 1.3) as 0.25 g. For detailed design, variations in seismicity within the EEC area would need to be considered.

4.54 Under the design earthquake loading an island may suffer some damage but its integrity should not be affected. The plant itself may also suffer damage but its safety should not be reduced.

4.55   Under operating conditions the plant may be subject to motions caused by sea action on the island. If these motions become excessive they will interfere with the operation of the plant and in the extreme will cause damage. A limit of 0.02 g acceleration in any direction has been taken as the requirement during operation of the plant.

Sea bed

4.56   The sea bed properties will play an important part in the selection of the optimum site and will influence the economic comparison between different types of island. As this study is not site specific, the influence of the sea bed properties can only be discussed generally.

4.57   The sea bed properties will govern the settlement, stability and feasibility of a given type of island. The influence of cyclic loading and earthquake motions may have a large effect on these factors.

4.58   The sea bed in the North Sea and English Channel is predominantly sand or gravel. The main areas of silt/clay are in the central North Sea. Rock occurs at the seabed along the western margins of the land masses facing the Atlantic Ocean. In some areas strata of clay are interbedded with the sands. While some of these clay strata are relatively strong others are not. Even those islands with low bearing pressures may not be feasible in the latter areas.

4.59   Unlike the North Sea, the English Channel and the eastern Bay of Biscay which are generally less than 200 m deep, the Mediterranean is generally very deep (2,000 - 3,000 m) with a narrow continental shelf. However, the Adriatic is shallow being generally less than 150 m deep. The geology of the narrow continental shelf and the Adriatic is very varied. Silty clay sediments dominate the Rhone delta region. The margins of the Adriatic are generally sandy-silts with finer sediments further from the coast.

4.60   Any prospective site will have to be investigated very thoroughly by boring and sampling down to 150 m or more for those islands with high bearing pressures.

4.61   Testing will include the determination of the strength and consolidation characteristics of the various soils encountered. From these the stability and settlements can be calculated.

4.62   The susceptibility of any granular material to liquifaction must be determined. In general the dense sands are not subject to liquifaction under earthquake motion whereas loose sands will be. It may be possible to excavate or compact the loose sands provided they are not too thick. However, the economic penalty such work will incur may render such sites unattractive. It should be noted that hydraulically placed sands are not in general subject to liquifaction as they compact to a fairly dense material.

Nuclear power station

Power plant layout

4.63 Our terms of reference and subsequent clarification require us to consider a 6 hectare site for the main components of the power plant and a total area of up to 15 ha for each complete 1,250 MW PWR unit.

4.64 The layout of the PWR was set out in a note on basic parameters which was handed to us at the 13th meeting of the Working Party. It is attached as Appendix B.

Power plant weights

4.65 The unit loadings were set out in the note on basic parameters - see Appendix B.

4.66 Total weights of the individual parts of the plant are required for calculating displacements of floating islands. The total weights cannot be calculated directly from the unit loadings given. Following discussions with the CEGB and further data from EDF, the total design weights of a 1,250 MW power plant for use in the study were established. These are given in Table 2.

Table 2 Total design weights for 1,250 MW plant

	Tonnes
Reactor building	170,000
Turbine hall	120,000
Fuel building	25,000
Nuclear auxiliary building	40,000
Control room building	25,000
Electrical and safety auxiliaries building	35,000
Transformers	3,000
Gas treatment buildings	10,000
Diesel generators	2,000
Total plant weight	430,000 tonnes

Access

General

4.67   Access to the island will be required for:

> construction materials and plant
> personnel during construction and operation
> nuclear fuel
> operational equipment and stores
> replacing plant during maintenance
> general maintenance.

4.68   It is not possible to establish specific criteria for all these requirements because too many variables are involved. However, an indication of the requirements is given in the following sections.

Construction materials and plant

4.69   The quantities of materials and plant will depend on the type of island. For islands that are constructed offshore, sea transport using a work harbour will be required until such time as a fixed access (if any) such as a road embankment is built. The construction of work harbours and fixed accesses are discussed in Section 7.

Personnel

4.70   The number of personnel requiring transportation to the island is discussed in Section 4.76 below.

Nuclear fuel

4.71   The weight of the flask required to contain the spent fuel is approximately 100 t. On average one movement every three weeks can be expected for a 2,500 MW development.

Operational equipment and stores

4.72   The requirements for operational equipment and stores are modest. Fuel oil and fresh water are the main liquids. A small tank facility and tanker berth will be required for these. The volume of stores will require a small wharf if a fixed access is not provided.

Replacing plant

4.73   The largest item of plant that may need to be removed from the island is the rotor from the generator. This weighs about 400 t. The main transformers will weigh less than this.

General maintenance

4.74   If a fixed access is not provided materials and plant for general maintenance can be handled by a small roll-on, roll-off facility. Such a facility can also handle the transportation of replacement plant and stores. For design purposes a 3,000 tonne ro-ro vessel adapted for handling large loads has been assumed.

Availability

4.75   Access to the island for personnel and emergency equipment should be possible in all weather conditions.

Construction and operation personnel

General

4.76   The number of construction personnel will depend on the type of island. For islands that are constructed and fitted out in wet and dry basins, construction personnel can be housed on land; some may be locally recruited and not require housing. For islands that are constructed in situ the problem of transporting personnel to the site has to be overcome. For all types of island the operation personnel have to be transported to the island.

4.77   In order to assess the transportation requirements for both the construction and operation periods, the number of personnel has to be determined. The transportation facilities during the construction period have to be capable of handling the peak number of personnel during the period. During the operation period the numbers will remain fairly constant.

Construction personnel - transport

4.78   The construction personnel can be split into those building the island and those building the power plant. The construction of a fill island will be a highly mechanised operation and the total labour force is unlikely to exceed 500. The maximum number of any shift will be about 300. The number working on the island construction will be running down as the number working on plant construction increases. The peak number of personnel on plant construction may be about 2,000 for a 2,500 MW development - this peak will not be achieved until after the island is substantially complete. For a 5,000 MW development the peak number may reach 3,700. Allowing for double shift working the peak number of personnel on any shift will be between 1,200 and 2,200. The transportation facilities should therefore be capable of handling these numbers of people.

Operation personnel - transport

4.79   The numbers required for operation will be much less than those for construction. The manning levels will vary considerably between various authorities and are therefore only approximate. The minimum manning levels have been taken as follows:-

Development	2,500 MW	5,000 MW	10,000 MW
Shift staff	90	180	360
Day staff	240	380	720
Total	330	560	1,080

4.80   The maximum transportation has been taken as being equivalent to transporting all the day staff at the same time. Because a proportion of the day staff are purely administrative it should be possible to establish a shore facility where such staff can work. This will relieve the pressure on the permanent transportation system and reduce costs.

Accommodation

4.81   If there is no permanent access during the construction period it may reduce overall costs to build a camp on the island to accommodate the workforce. The cost of the camp can be offset against the reduced cost of transportation and the benefits of less travelling time for the construction personnel.

4.82   During the operational period it will be necessary to provide accommodation for the shift staff. In addition, accommodation of a lower standard may be required to house day staff in the event of severe weather conditions.

Transmission links

4.83   The transmission link to the shore has been assumed to be at 400 kV using 3 phase, 2,500 MW circuits. The cables will be buried in trenches in the sea bed. For 2,500 MW and 5,000 MW developments full standby capacity has been assumed. For the 10,000 MW developed it has been assumed that the additional 5,000 MW of development will share the standby capacity of the first 5,000 MW development. The use of D.C. transmission has not been examined but would form part of any subsequent design study.

A. Floating island

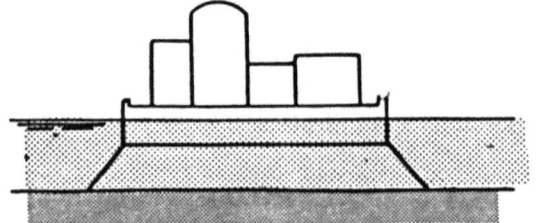

B. Integral floating island

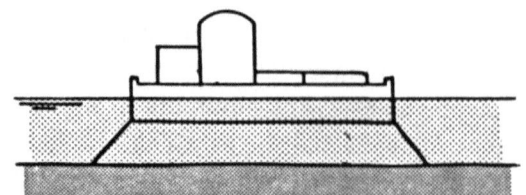

C. Semisubmersible island

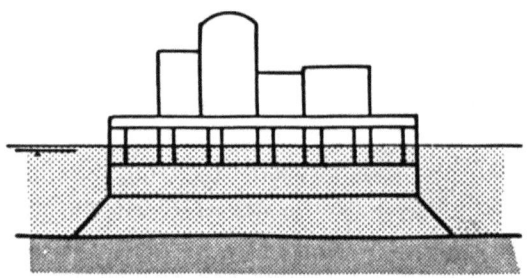

D. Gravity platform

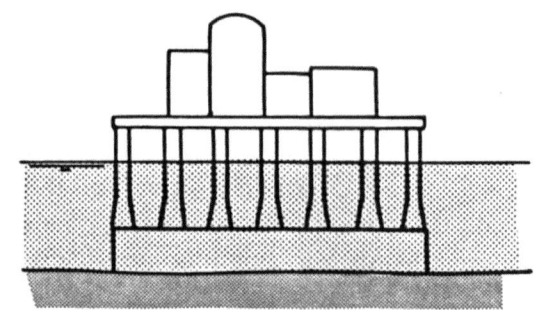

E. Legged platform

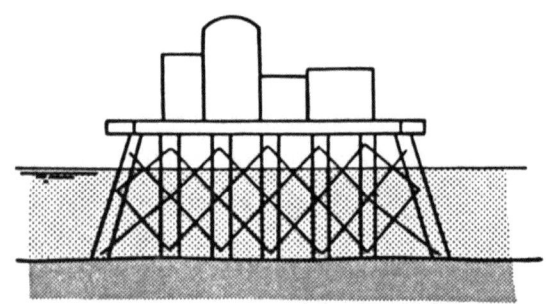

F. Stayed tower

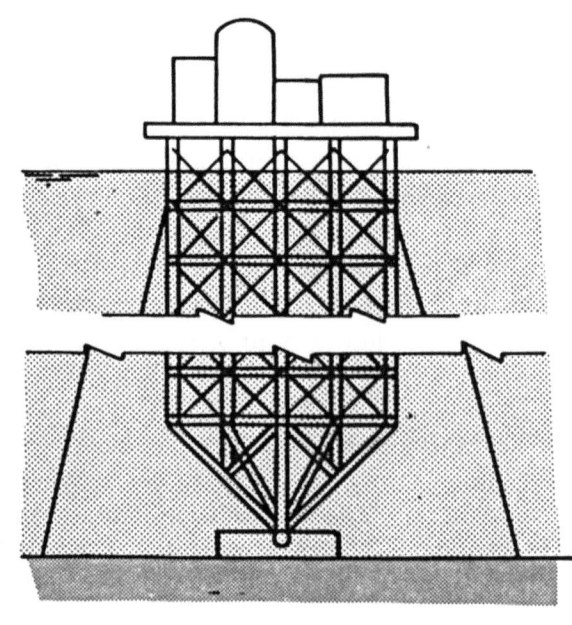

G. Caisson island

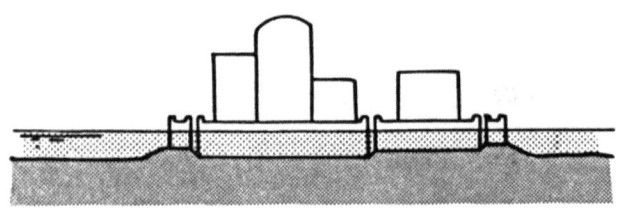

I. Piled island

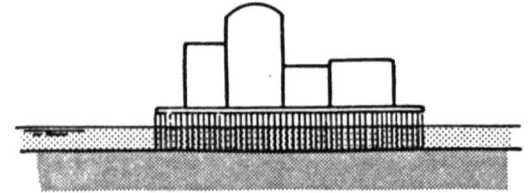

H. Subsea island

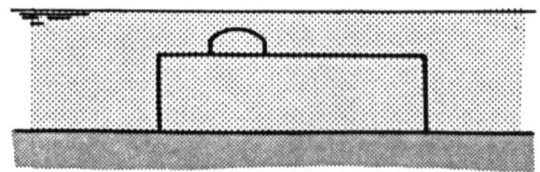

Island concepts – Sheet 1

Drawing 2

5. ISLAND CONCEPTS

Introduction

5.1   A broad conceptual review has been carried out to identify the types of island that can be considered for supporting a nuclear power plant. These types are given in Section 5.2. Their advantages and disadvantages are discussed in Sections 5.3 – 5.38 and the most promising types for outline design are selected in Sections 5.39 onwards.

Types of island

5.2   The types of island identified are listed below and illustrated on Drawings 2 and 3.

Floating islands

Type			
	A.	"floating island"	– Raft of cellular steel or reinforced concrete construction with the power plant on top of it.
	B.	"integral floating island"	– Raft of cellular steel or concrete construction with the power plant forming an integral part of the raft.
	C.	"semisubmersible island"	– Semisubmersible vessel of cellular steel, reinforced concrete construction or composite construction.

Fixed islands

Type			
	D.	"gravity platform"	– Concrete gravity structure founded on the sea bed with a fixed platform above sea level.
	E.	"legged platform"	– Legged steel structure (jacket) pinned to the sea bed with a platform above sea level.
	F.	"stayed tower"	– Lattice tower structure, attached to the sea bed, with a platform above the sea bed, the whole being held vertical by stays.
	G.	"caisson island"	– Concrete caissons found on the sea bed with their tops forming a platform above sea level.
	H.	"subsea island"	– Concrete caissons enclosing the power station and founded on the sea bed; the structures being below the sea surface.
	I.	"piled island"	– A reinforced concrete deck supported above sea level on bearing piles driven into the seabed.

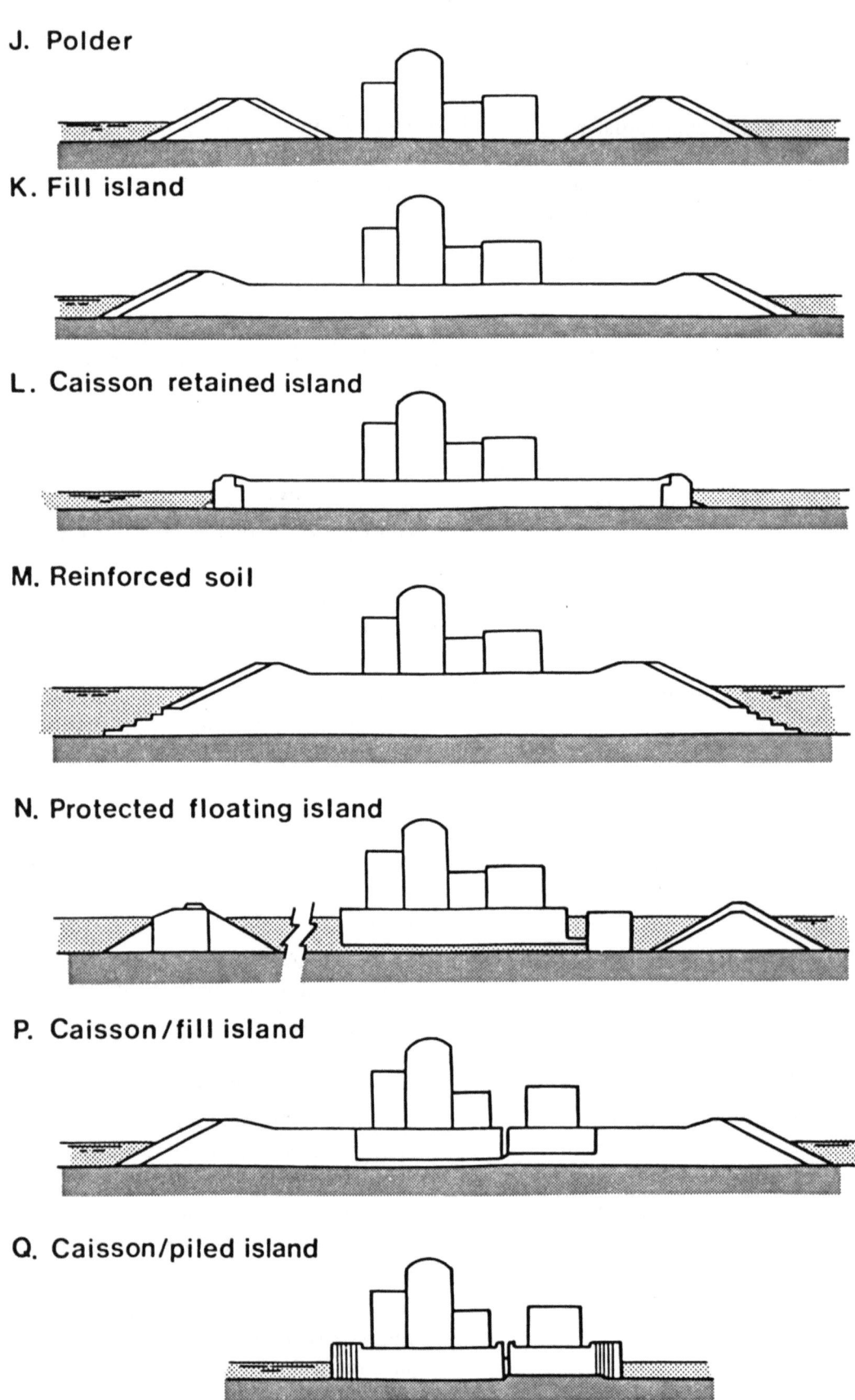

Island concepts – Sheet 2   Drawing 3

Fill islands

Type	J.	"polder"	–	A polder type reclamation with a fill embankment with suitable sea protection.
	K.	"fill island"	–	A fill type reclamation with a sand fill and suitable sea protection.
	L.	"caisson retained island"	–	A fill type reclamation with sand fill retained and protected by a perimeter of caissons.
	M.	"reinforced soil"	–	A fill type reclamation with sand fill retained and protected by a reinforced soil skin.

Composite islands

Type	N.	"protected floating island"	–	A floating island (A) within a protective breakwater.
	O.	"caisson/fill island"	–	A composite island with the main plant components on caissons and auxiliary plant on a fill island.
	P.	"caisson/piled"	–	Concrete caissons supporting the main components with the remainder of the island on piles.

Features of island types

Floating islands - Types A, B and C

Advantages and disadvantages

5.3   Some of the advantages of floating islands are that they:

   (a)   lend themselves to prefabrication and mass production

   (b)   are relatively easy to remove at the end of their operational life

   (c)   are insensitive to poor ground conditions except for anchorage.

   (d)   can be sited in very deep water provided the anchorage and protection systems can be made adequate.

5.4   Some of the disadvantages of floating islands are that they:

   (e)   are floating and can therefore sink

   (f)   are vulnerable to collision and sabotage

(g) are subject to movement, particularly tilt, and vibration from wave and wind loadings

(h) require a deep water basin for construction and float out but see (a)

(j) have moorings which are vulnerable to corrosion and sabotage and are difficult to maintain

(k) have poor access in bad weather

(l) are unsuitable for siting category I because of their draught

5.5   The floating island, is a simple concept and its construction is well within the bounds of current technology. If it is constructed in one raft its size will make it difficult to handle and vulnerable to long waves. The power plant on the deck is exposed to collision and corrosion.

5.6   If the items of plant are located on several small rafts, some of the above problems are overcome. In addition the size of the shore construction facilities can be reduced. However, problems of connecting units and of differential movement are introduced. The smaller units are also more susceptible to movement under wave attack.

5.7   The integral floating island, type B, is a more radical solution but requires modifications to the basic layout of the power station. It could produce cost savings. The reactor is housed partially within the raft which provides protection for it. Construction on multiple floors within the raft increases the complexity but could reduce the overall plant area.

5.8   The semisubmersible island, type C, is a more complex structure. The effect of waves is reduced at the expense of extra height above sea level. Structures of this type have been used in the North Sea although they are smaller than those required for a nuclear power station.

Design factors

5.9   Particular study is necessary of wave loadings and the typical long waves that could be found at the various siting categories. The distribution of the nuclear plant within the raft and the wave and wind forces are critical for the loading on the submerged structure. The overall stability of the structure, dynamic loadings and resonance require careful analysis.

5.10   The mooring system is an important aspect of the design. Various types are possible including catenary mooring cables, tensioned cables, chains, struts, tensioned "legs" and combinations of these. The anchorages can be on the seabed or on towers, caissons or breakwaters.

5.11   The towing of the island from its final construction basin to its permanent location may present severe difficulties if tidal or other currents are strong.

Vulnerability

5.12   Floating islands are inherently more vulnerable than fixed islands. They can be designed to withstand natural conditions of extreme severity which have a very small exceedance value. However, statistically the risk of sinking is still finite. In addition the risk from man-made hazard is also present. Ship collision and sabotage fall into this category. Collision can to some extent be designed against but sabotage is always a possibility and cannot be entirely ruled out. For this reason we consider that floating islands should only be used if they are designed to withstand being sunk without causing a major catastrophe. This may mean that the floating island will have to be moored in water not much deeper than its draught. In this case, large tidal ranges could cause problems.

5.13   Even if the concept of the island being allowed to float in deep water is accepted, the problems of protection from collision and waves are difficult. Tethered floating ship arresters for use in the open sea against very large crude carriers have not yet been developed. Floating and air curtain breakwaters are not suitable for the exposed locations being considered. If the risk of collision is to be minimized and exposed conditions are considered, an armoured rockfill breakwater around the floating island may be necessary. Such a breakwater will be expensive in deeper waters.

Suitability

5.14   The use of floating islands for nuclear power stations poses several fundamental problems of the acceptability of risks and hazards. These problems could severely restrict the use of floating islands.

5.15   Floating islands cannot be used in siting category A because the water is too shallow for the island to float.

Fixed islands - types D to I

Advantages and disadvantages

5.16   Some of the advantages of fixed islands are that they:

(a)   lend themselves to prefabrication and mass production

(b)   are more stable than floating islands.

5.17   Some of the disadvantages of fixed islands are that they:

(c)   are vulnerable to collision and sabotage

(d) are subject to vibration from wave and wind loadings

(e) require a construction basin with very deep water

(f) require a prepared foundation on the seabed

(g) are vulnerable to corrosion and are difficult to maintain

(h) have poor access in bad weather.

(j) are unsuitable for siting categories I and II because they require too great a depth of water. An exception is type I.

General

5.18   The gravity platform, type D, is similar to some of the structures used in the development of the North Sea oilfields. The platform above sea level is supported on columns which rise from a broad base founded on the seabed. The base of the structure can be used for oil storage. When used for a nuclear plant this configuration is less suitable unless a radical rearrangement of the plant is possible. Such a rearrangement, which might raise the problems of safety regulations, is beyond the scope of this report and is not considered further. The loading from a nuclear power station is larger than for any existing designs and several islands would probably be required to support a 2500 MW development. This type is likely to become more economic than type G as water depth increases.

5.19   The legged platform, type E, is similar to many of the steel structures used in the North Sea. The loading from a nuclear power station is larger than for any existing design. Several individual islands will probably be required. These structures have usually been constructed in dry docks on their sides, floated into location using a barge and then turned upright. This method precludes any degree of fitting out of the power plant in the dry dock. An alternative is to use a jack-up platform founded on the same type of legs. Pinning these structures to the seabed to resist overturning in deep water involves difficult piling. However, the need for a large level area for the foundations is avoided. The protection of the splash zone on a steel structure against corrosion for a design life of 80 years is a major problem.

5.20   The stayed tower, type F, is a relatively new concept that has recently been considered for use in oil field development. It is a development of the legged platform. The weight of the platform is supported on a lattice tower pinned to the seabed but the stability is achieved by restraining the top of the tower with cables anchored to the sea bed. Because the weight of a nuclear power plant is much greater than oil field installations, the foundation for the base of the tower would have to be bedrock and the base of the tower would be so massive that the concept would not be feasible. A stayed tower also suffers from the major problem of corrosion for the 80 years design life. The restraint from the stays is unlikely to be enough to stop excessive tilt of the structure.

5.21 The caisson island, type G, is a simple solution which is within existing technology. The slab sides present less surface for corrosive attack but are subject to heavier loadings from waves. The power plant can be built into the structure and a large portion of the installation could be done before towing it onto location. The advantages of this type may, perhaps, be best exploited in a composite island.

5.22 The subsea island, type H, being beneath the surface, is environmentally attractive. However, the problems of safety and access are severe and although such a development may eventually be proved feasible, it is unlikely to be so for many years.

5.23 Unlike the other fixed islands, the piled island, type I is suitable for shallow water sites. The 'legs' supporting the concrete deck are formed by driven piles. Because the platform is constructed in-situ, cross bracing can only readily be provided above sea level. Tying the piles together underwater would be a complicated and costly operation. Without lateral support beneath the sea surface the height of a piled island is restricted by buckling considerations. Under the main items of nuclear plant loadings are high, which increases the tendency to buckling. The nuclear plant has to be constructed on the island and not at a shore facility unlike the other fixed islands.

Design factors

5.24 The wave forces on the supports of the legged structures and the scour at the base of gravity structures require study. The dynamic response and vibration of these structures also need to be considered. The problems of collision are similar to those for floating structures. Foundation conditions play a major part not only in the design of the structure but in deciding on its feasibility. For steel structures corrosion and fatigue under cyclic loading are a major design problem.

Vulnerability

5.25 Fixed islands, like floating islands, are inherently more vulnerable than fill islands. The risks from ship collision and sabotage are a major problem. Steel legged structures are more vulnerable to sabotage than concrete gravity platforms.

Suitability

5.26 The fixed island is unlikely to be economic for siting categories I or II but may be economic for the deep, exposed waters of category III.

Fill islands - types J to M

Advantages and disadvantages

5.27 Some of the advantages of the fill islands are that they:

(a) are not subject to movement and vibration under wave and wind loadings

(b) are the least vulnerable to collision and sabotage

(c) do not require deep water construction basins (caissons excepted)

(d) are less subject to corrosion and are easier to maintain

(e) have better access in bad weather.

Some of the disadvantages of fill islands are that they:

(f) have to be built in-situ

(g) require large quantities of rock

(h) have to be partially built before power station construction can start

(j) may be difficult to demolish.

General

5.28   The polder, type J, is simple and has been extensively used for land reclamation, particularly in the Netherlands. In principle the main advantage is that less fill material is required because the area within the enclosing embankment would be left at seabed level. This advantage can only be fully utilized if an area of several square kilometers is being reclaimed. The fill saved for smaller enclosures is unlikely to offset the additional capital and running costs of maintaining the area free of water. A major disadvantage of the polder is that damage to the embankment could lead to the power station being flooded.

5.29   The fill island, type K, is also simple and has been used on a small scale. This form of construction can be extended to any required area of island. The main element of the island is the sea protection against waves and tides. This can take many forms including artifical beaches of sand or gravel, layers of graded rock, precast concrete blocks of regular or random orientation, asphalt revetments, concrete blockwork walls, caissons and sheet piles. This type of island is the most resistant to collision. Suitable fill material must be available locally if costs are to be minimized. Because construction takes place offshore, weather conditions will have more influence on down time and construction costs. An advantage of the fill island is that cooling water recirculation is reduced by the larger bulk of the island.

5.30   The caisson retained island, type L, is a variation on the fill island where the perimeter of the fill is retained and protected by a continuous line of caissons. This type of construction offers advantages in locations subject to intermittent severe conditions as, once positioned, the caissons offer good protection to both the fill material being placed and the construction plant. Their disadvantage is that general filling cannot start until some of the caissons have been brought to the site and thus construction time may be longer for this type of island. In deep water it is possible to found the caissons on a prepared fill platform. However, scour at the toe of the caissons requires careful consideration.

5.31   The reinforced soil island, type M, is another variation on the fill island. In deep water the volumes of fill in the outer sloping faces of a conventional fill island are large. Although these volumes can be reduced by placing the fill within confining rockfill bunds, the rockfill required may be expensive or in short supply. The use of reinforced soil may allow the soil slope to be steepened considerably thus saving large quantities of fill. The technique involves the laying of layers of mesh in the fill thereby increasing the effective shear strength of the material. This allows steeper slopes to be achieved. In practice construction difficulties limit the steepness. This technique is currently being considered for building islands in the Arctic oil fields.

Design factors

5.32   The various slope protection methods and their relative construction and maintenance costs need to be studied. The settlement of the fill, its effect on the foundations of the power station, scour of the embankment and possible modification of the local patterns of sediment transport also need study.

Vulnerability

5.33   The fill islands are the least vulnerable of all the types of islands - they cannot sink or be turned over although a polder island could be flooded. Ship and aircraft collision are unlikely to affect the development any more than for a power station located on the sea shore.

Suitability

5.34   Fill islands will be suitable for all siting categories but will become more difficult to construct and possibly uneconomic as the water depth increases in siting category III.

Composite islands - types N to Q

Advantages and disadvantages

5.35   In a composite island some of the individual advantages of the various floating, fixed and fill islands can be combined. Various combinations will individually be best suited for different sites. Some of the more promising combinations are considered below.

General

5.36   The protected floating island, type N, combines the floating island's advantages of prefabrication and mass production with the greater security against collision and wave attack afforded by a fixed breakwater. However, the cost of the breakwater will rise with increasing water depth. Because the floating island itself cannot be used in shallow water, the protected floating island will only be feasible in the water depths of categories II and III. Costs will make it unattractive in the deeper parts of category III.

5.37   The caisson/fill island, type P, combines the advantages of the main power plant being prefabricated on caissons with the greater security of the fill island. The construction of the caissons and installation of the power plant (or part of it) can be carried out while the fill island is started. The caissons can then be towed into position and the fill island completed around them. The caisson/fill island can be used in category I if approach channels for the caissons can be dredged during filling of the island. The caisson fill island will become uneconomic in the deeper waters of category III.

5.38   The caisson/piled island, type Q, is similar in concept to the caisson/fill island, the replacement of filling by piling being the economic difference. The piling around the perimeter will be more vulnerable to collision than fill. Practical limits on pile sizes mean this type of island is not suitable in deep water.

Comparison of island types

5.39   The types of island are compared under specific key factors in table 3. The rating for a factor indicates whether the factor is a point in favour of (good rating) or against (bad rating) the particular type of island. The suitabilities of the types of island for the three siting categories are also indicated.

5.40   The key factors are split into three groups:

>technical acceptability
>economic factors
>construction factors

5.41   Technical acceptability is considered under five headings. If an island type rates badly under any of these headings its feasibility is in doubt. The economic factors indicate the sensitivity of island costs to site conditions. The construction factors indicate the degree of prefabrication that is possible both for the island and the plant and also the extent and difficulty of the shore installation that will be required.

Category I - Island suitability

5.42   The 10 m water depth available is about half that required for floating islands. Although both the approach channel and the site could be dredged this would be wasteful. Fixed islands types D to H require large water depths for positioning and are also unsuitable. Piled islands type I can be built in shallow water but are more vulnerable to collision damage and corrosion than fill islands. Fill islands can be constructed in shallow water or the tidal zone with equal ease. Composite islands such as caisson/fill or caisson/piled islands are possible but dredged approach channels will be required for the caissons and prefabricated plant units. We conclude that fill islands are the most suitable for category I.

TABLE 3 COMPARISON OF ISLAND TYPES

ISLAND TYPES

KEY FACTORS	A Floating Island	B Integral floating island	C Semisubmersible	D Gravity platform	E Legged platform	F Stayed tower	G Caisson island	H Subsea island	I Piled island	J Polder	K Fill island	L Caisson retained island	M Reinforced soil island	N Protected floating island	O Caisson/fill island	P Caisson/piled island

Technical acceptability

- Availability of required technology
- Vulnerability to collision
- Vulnerability to corrosion
- Ease of maintenance
- Dynamic stability

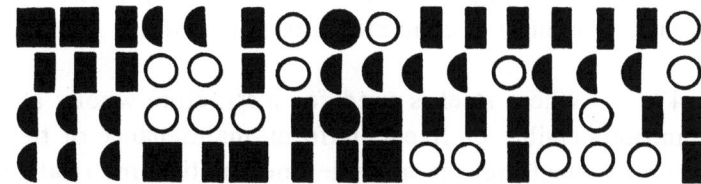

Economic factors

- Cost sensitivity to waves
- Cost sensitivity to currents
- Cost sensitivity to tidal range
- Cost sensitivity to foundations

Construction factors

- Prefabrication of island
- Prefabrication of plant
- Shore facilities

Suitability

- Category I
- Category II
- Category III <40 m deep
- >40 m deep

● ◐ ○ ▌ ■
good rating   bad rating

Category II - island suitability

5.43 The 10 to 20 m water depth available in category II could accommodate a floating or integral floating island but some dredging would probably be needed. The semisubmersible island will require greater water depths. The caisson island and piled island are a possibility. However, the other fixed islands require more than 20 m depth to be feasible. The fill island is suitable for this category as are the composite islands although some dredging could be required for the latter. A broader range of island types are thus suitable for category II.

Category III - island suitability

5.44 For the shallower water depths of category III the fill and composite islands are still feasible. However, for the deeper waters only floating or fixed islands are feasible.

Comparison of concrete and steel

5.45 In view of the required design life of 80 years and the generally hostile marine environment, we consider that concrete should be used instead of steel for all structural work below sea level or in the splash zone. Where this is not possible the feasibility of the island must be in doubt.

5.46 Both concrete and steel structures have been used for oil production platforms in the North Sea. This suggests that concrete and steel structures are competitive in these conditions. Proposals for oilfield developments in deeper water make use of both materials. However, the design life for an oil production platform is only about 20 years so that the greater durability of concrete is not important.

5.47 Where steel is used for a structure wholly above sea level such as the deck of a semi-submersible or fixed platform, maintenance of the structure is greatly simplified. In this case steel may be a viable construction material for the design life being considered.

5.48 The choice of a concrete deck instead of a steel deck is unlikely to produce much direct decrease in costs. The reduced weight of material needed for a steel deck being largely offset by the higher cost of steel compared with concrete. However, the structure as a whole is lighter with a steel deck than a concrete one. Less buoyancy is therefore needed and so a steel deck may produce a saving by reducing the size and cost of the submerged part of a structure.

5.49 For a semisubmersible the weight of the deck is only about 20% of the total displacement of the island. If substitution of steel for concrete halves the weight of the deck, a saving of about 10% in the overall cost of the island may be achieved. The potential savings for fixed structures are lower because the size of the submerged part of the structure is determined largely by foundation loadings. In this case flotation is only a secondary function of the structure.

5.50 The feasibility of steel or composite deck construction would need to be investigated further before it could be recommended for the long design lives under consideration. In any case the cost savings from the use of these materials is likely to be only a small proportion of the total cost.

5.51   In this Report the outline designs for floating and fixed island are based on the use of concrete rather than steel.

Conclusions

5.52   There are many types of island that could support nuclear power station development. However, the choice of types available in shallow or deep water is limited.

5.53   Because there are reasonable alternatives for them, the following types of island that have technical disadvantages are not considered any further:

floating islands of steel	(type A - part)
semisubmersible islands of steel	(type C - part)
legged platforms	(type E)
stayed towers	(type F)
piled island	(type I)
polders	(type J)

5.54   In addition the following types of island that are at or beyond the limits of existing technology are not considered further:

integral floating islands	(type B)
subsea islands	(type H)
reinforced soil islands	(type M)

5.55   The most attractive types which are considered further in the section on outline designs, are:-

floating island (concrete)	(type A-part)
semisubmersible island (concrete)	(type C - part)
gravity platform	(type D)
caisson island	(type G)
fill island	(type K)
caisson retained island	(type L)
composite islands	(types N, P, Q)

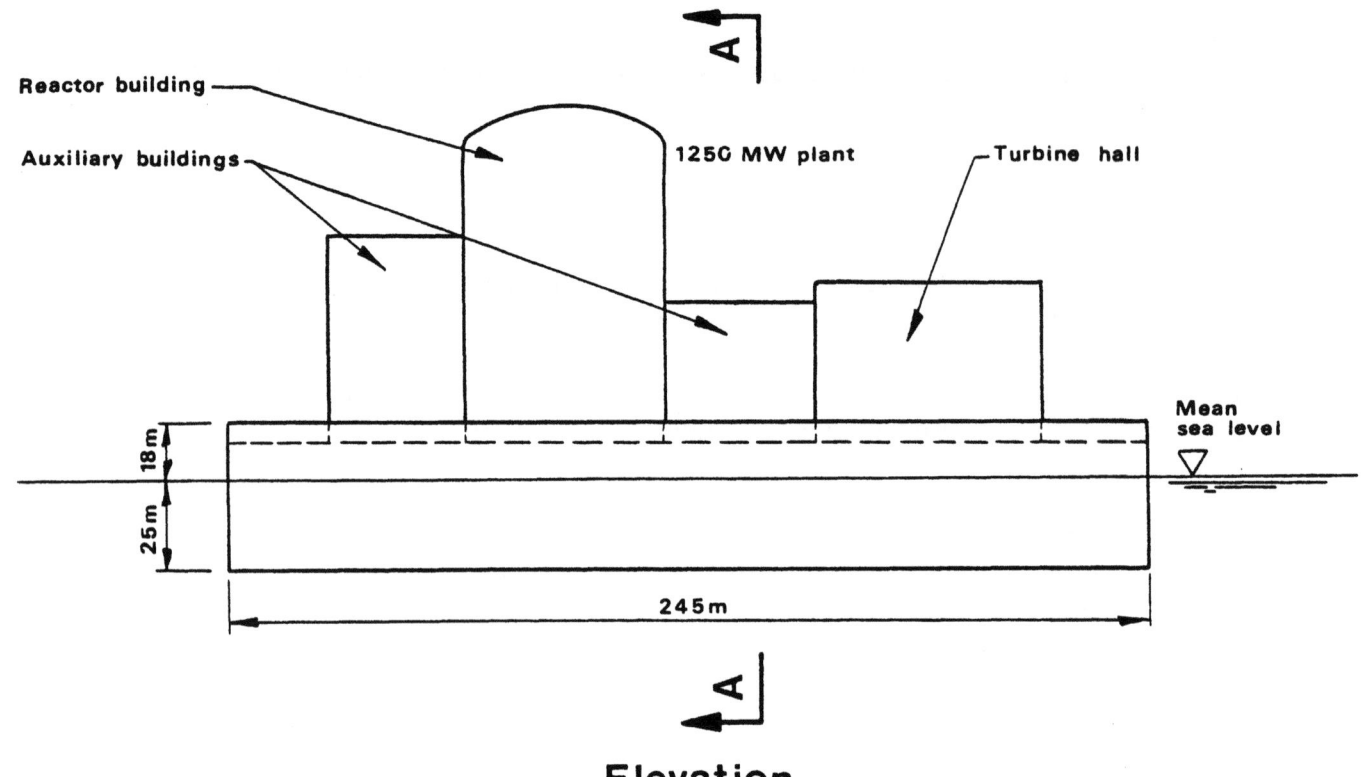

Elevation

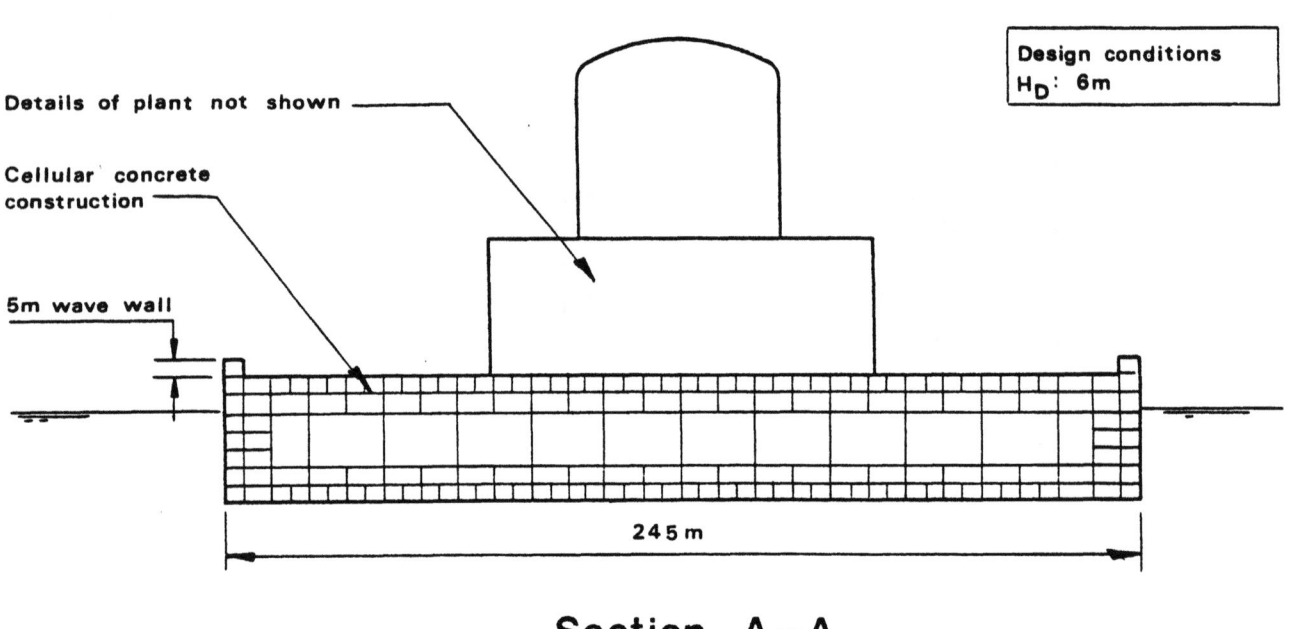

Section A-A
Scale 1:2000

Floating island

Drawing 4

6. OUTLINE DESIGNS

Introduction

6.1   The review of island concepts has identified the most suitable types of island for the three siting categories. In this section outline designs are described for these types of island and the design problems discussed. It has been necessary to qualify the feasibility of some of the types - more detailed design or advances in technology may remove some of the qualifications.

6.2   There are problems such as construction basins and tow-out that are common to several of the types of island and these are reviewed in the section on construction methods - see Section 7. The various ancillary requirements such as access, harbours and cooling water are discussed in Section 6.102 onwards.

Floating island

General

6.3   As discussed in Section 5, outline designs have been prepared for concrete structures. Although the raft type of floating island is unsuited to severe wave conditions, design wave heights (H_D) in the range 3-12 m have been considered.

Layout

6.4   The layout of the plant in plan is as shown on the sketch in the draft information note (Appendix A). The main plant items for the single 1250 MW unit occupy 6 hectares.

6.5   Improved stability can be achieved if the raft is made square, 245 x 245 m. The plant can be accommodated in this square raft without rearrangement. However, a more compact arrangement is achieved by turning the turbine hall through 90°. The improved weight distribution given by this layout allows a uniform cross-section of raft to be adopted. Small adjustments of trim can be made by altering the distribution of the ballast.

6.6   A 2,500 MW development will consist of two main rafts, each housing a 1,250 MW unit, and a series of smaller rafts for the ancillary equipment.

Design concept

6.7   A cubical cellular concrete arrangement has been used for the design. The minimum cell size taken is a 5 m cube. These cubes form the outer skin of the island. The internal cells are larger and can be up to 20 m cube depending on deck loadings and access requirements. The cellular construction has good strength when subject to bending caused by long waves. The walls of the cells form continuous diaphragms running along and across the island.

6.8   Cylindrical cellular units have been considered. These units in general give thinner wall thickness for a given external loading. However, they would have to be stiffened by additional longitudinal and transverse diaphragms to resist the bending caused by long waves. Thus the simpler cubical cells which naturally form such diaphragms appear the better type.

Draught and freeboard

6.9   The freeboard needed above mean sea level depends on the design wave. For an island free to rise and fall with the tide a freeboard of 3 x H_D has been taken.

6.10   The main deck level of the raft need not be at or above the freeboard level. The upper part of the wave protection can be provided by a wave wall rising above the main deck level.

6.11   The island has been designed to float with the design freeboard when 10% of the structure is flooded. Under normal operating conditions the island would be maintained at its design freeboard by ballasting. The outline design is shown on Drawing 4.

6.12   A 1250 MW power plant weighs 430,000. The dimensions of a 245 x 245 m raft to carry this plant with the required freeboard for a range of design wave heights are:

H_D	Weight of concrete in island (Mt)	Displacement when ballasted	Draught (m)
3	0.59	1.20	20
6	0.92	1.50	25
12	1.71	2.40	40

Dynamic stability

6.13   The general factors affecting the dynamic behaviour of structures are discussed in Appendix K. A brief study of the behaviour of the floating island under wave action follows.

6.14   A floating island has to resist large forces generated by waves. These forces are caused by the wave action against the external surface area and by the changes in distribution of buoyancy as waves pass the island.

6.15   A small vertical displacement of the raft produces a large restoring force. Similarly because the plan area of the raft at the water line is large a small rotation about a horizontal axis produces a substantial restoring moment. A floating island therefore behaves as a stiff system when subjected to heave, pitch and roll. In these modes the natural period of a floating island is about 10 seconds. The period does not change much over the range of island weights being considered. Because the natural period of oscillation in these modes is in the same range as the periods of the incident waves, severe motion is likely to result if the island is unrestrained.

Mooring

6.16   A mooring system is required to keep the island at its location and resist quasi-steady forces such as winds and currents. In addition the dynamic stability analysis shows that the mooring must also resist rapidly varying loads if the motion of the island is to be kept within operational limits. As well as controlling the heave, pitch and roll of the island the moorings must also prevent undue surge, sway and yaw.

6.17   To limit the acceleration of a floating island during operation to 0.02 g (see Section 4.55) requires mooring forces comparable with the forces imposed by wave action. To prevent high initial accelerations the moorings must provide virtually the full restraining force before the island is displaced. Any initial slack would allow a high acceleration followed by deceleration as the moorings become taut. We therefore conclude that the mooring system needs to be pre-loaded.

6.18   The table below shows the approximate horizontal forces produced by a non-breaking reflected wave acting on one side of an island.

H_D (m)	Total force due to reflected wave on 245 m side (KN)
3	330×10^3
6	910×10^3
12	3000×10^3

6.19   Similar vertical forces are produced by changes in buoyancy due to long waves.

6.20   If moorings are used to limit the vertical movement of a floating island, the tidal range becomes important. Unless the length of the mooring lines or the ballast in the island can be adjusted in phase with the mean sea level, uplift forces on the raft will vary with the state of the tide. The moorings must be taut at low tide, and thus very large forces will be introduced into the moorings by tidal change - 600×10^3 kN per metre rise for a 6 ha island.

6.21   A further consideration is that the height of the raft above mean sea level needs to be increased, to provide adequate freeboard against wave attack at high tide.

Elevation

Section A–A

Scale 1: 2000

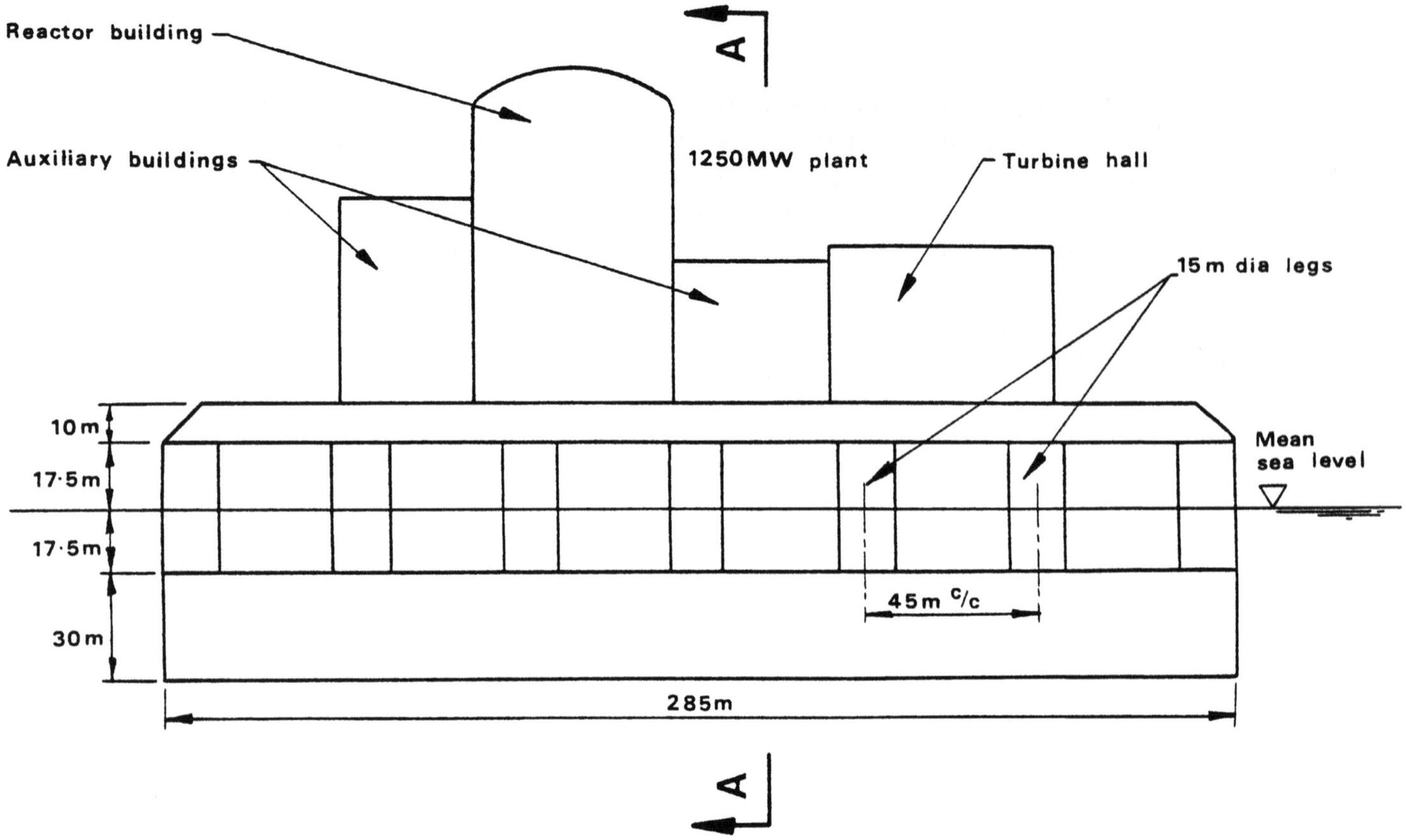

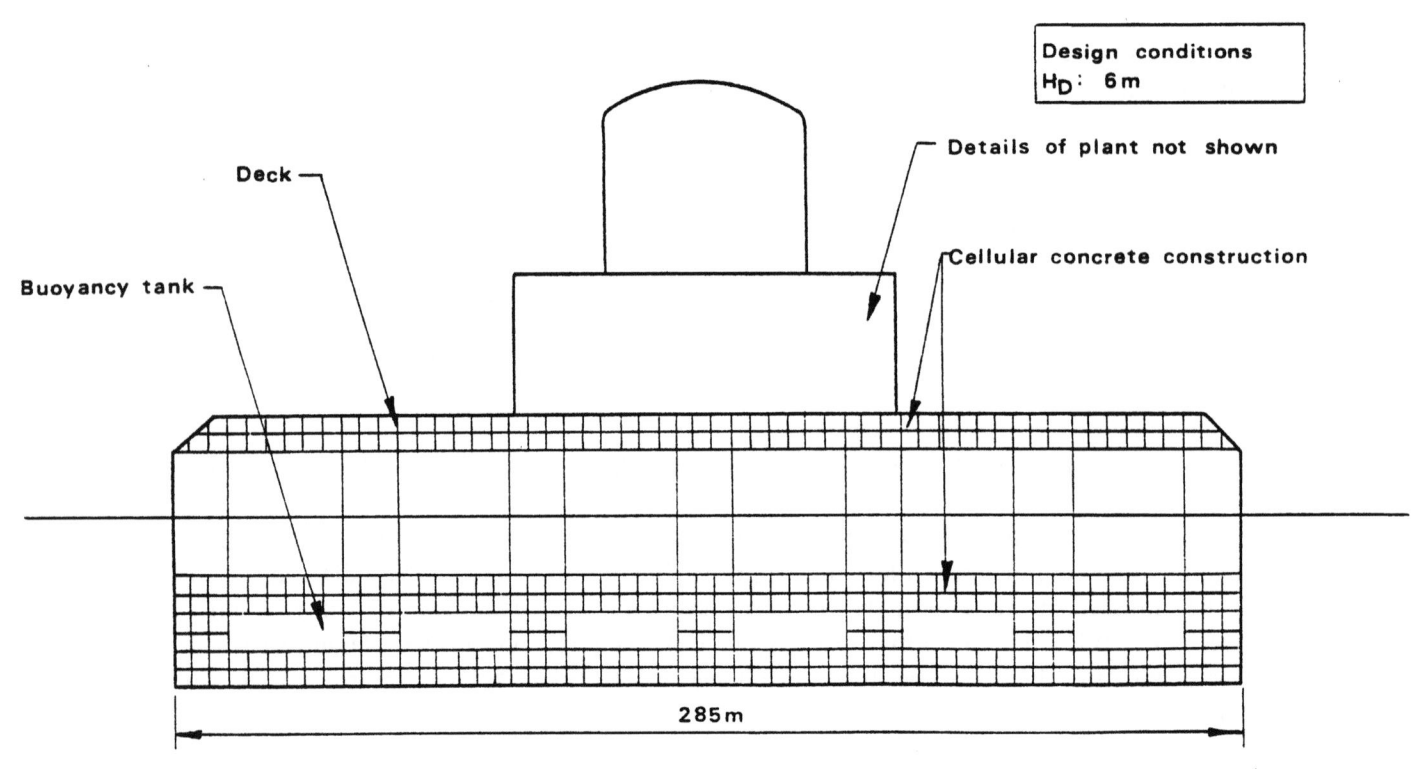

Semisubmersible island

Drawing 5

6.22   To resist wave forces and the effects of tides the moorings of a floating island need to provide large forces. Arrangements of mooring lines that might be used to fix a floating island have not been considered in detail. The maximum working load of existing cables and chains is about 5000 kN. Several hundred such mooring lines would be needed to anchor a floating island in exposed conditions. Existing moorings of steel wire rope or chain have typical working lives of about three years.

6.23   Loads of 5000 kN are in excess of the capacity of existing ship-type anchors. Special anchor blocks or piled fixings would be needed to connect the mooring lines to the sea bed.

6.24   In deep water, construction of the seabed anchorages becomes difficult. Also as the water depth increases an increasing proportion of the mooring lines strength is needed just to lift the weight of the cable or chain off the sea bed.

6.25   The number of mooring lines that can be connected to the raft of a floating island will be limited by the size of the raft. Anchorages may also need to be distributed over the seabed to avoid congestion and local concentrations of load.

6.26   Mooring technology is continually improving and mooring lines and anchorages are being developed for increasingly large loads. New materials are also under consideration which should increase the life of moorings. However, with current technology we consider that to moor a floating nuclear power plant at an exposed site and limit its motion enough to permit plant operation is not feasible.

Summary

6.27   The basic design concept of cellular concrete construction is simple in principle. The draught of the island is large and will cause construction problems. The main disadvantage of the floating island is that its natural period is of the same order as the high energy waves in the design storm. To restrict the motion of the island,, moorings capable of taking very large forces are required. We do not consider the unprotected floating raft type nuclear island is feasible.

Semisubmersible island

General

6.28   Concrete structures have been selected in preference to steel for the reasons given in Section 5.

6.29   The semisubmersible is better suited than the raft type of island for exposed locations. However for comparison with other designs we have considered design wave heights (H_D) in the range 3 - 12 m.

Layout

6.30   The plant layout is the same as for the floating island - a deck area of 6 ha has been provided.

6.31   The stability of a semi-submersible is also improved if the plant is based on a square deck 245 x 245 m, rather than a rectangular one.

Design concept

6.32   The semisubmersible island can be considered as three elements namely:

>   deck
>   supporting legs
>   buoyancy tank.

6.33   The deck has to support the various parts of the power plant and distribute the loading into the supporting legs. It has to be deep in order to span between the legs. A 10 m deep section has been adopted. The soffit of the deck must be above the top of the highest wave if wave forces are to be minimised. The deck is of cellular concrete construction.

6.34   The supporting legs must transfer the plant loadings to the buoyancy chamber and be able to resist bending due to wave loadings. The amount of water displaced by the legs when the semisubmersible is rolling or pitching determines the resistance of the structure to overturning. If the cross-sectional area of the legs is too small, insufficient restoring moment will be mobilised and the structure will be unstable.

6.35   Stability has been the overriding criterion in the choice of leg diameters and spacings. A total of 49 no, 15 m diameter legs at 45 m centres have been selected.

6.36   The buoyancy tank is conctructed on the same principles as the raft type floating island. The basic unit is a 5 m cubical cell.

Draught and freeboard

6.37   The height of the deck soffit above the waterline depends on the design wave. The maximum wave height is 2.1 x H_D so that the crest of this wave is 1.05 x H_D above stillwater level. Unlike a floating island where reflected waves of a greater height than the incident wave will be formed, little heightening of the waves should occur at a semisubmersible. We have taken a deck soffit level 5 metres above the highest wave.

6.38   The island has been designed to float with the design freeboard when 10% of the buoyancy tank is flooded.

6.39   To provide this reserve of buoyancy and to allow a sufficiently wide leg spacing for roll stability a buoyancy tank of 285 x 285 m has been chosen. A deck of similar area has been provided but only the central area of 245 x 245 m is occupied by the plant. Outside this region a lighter deck structure has been provided to give support at the top of the legs. The outline design is shown on Drawing 5.

6.40   The weights and draughts of semisubmersible islands for different design wave heights are:

H_D (m)	Weight of concrete in island (Mt)	Displacement when ballasted (Mt)	Draught when ballasted (m)
3	1.44	2.16	37
6	1.57	2.35	48
12	1.78	2.57	60

Dynamic stability

6.41   A semisubmersible has a smaller water surface area than a floating island. Thus smaller restoring moments and forces are produced when it is displaced from a position of equilibrium so it behaves as a less stiff system. The natural period for heave is about 30 secs while the periods for roll and pitch are in excess of 1 minute. The periods are similar for the range of designs being considered.

6.42   The natural periods are greater than the periods at which most of the wave energy occurs. However, although the semisubmersible is some way from resonance, the amplitudes of movement will still be large at the commonly occuring wave periods because the restoring forces are relatively small. The mooring system will be required to keep movement within operational limits as well as keeping the island on location.

6.43   A semisubmersible experiences forces due to wave diffraction about the legs of the platform. Hydrostatic forces act on the submerged buoyancy chambers as pressures fluctuate beneath the waves. The absence of any continuous surface at the water level avoids the high forces due to wave reflection that occur on a floating island. However, the semisubmersible has a larger submerged cross-section and this partly offsets the reduction particularly at lower wave heights.

6.44   The approximate horizontal forces acting on a semisubmersible over a range of wave heights are:

H_D (m)	Horizontal wave force (kN)
3	300×10^3
6	600×10^3
12	1200×10^3

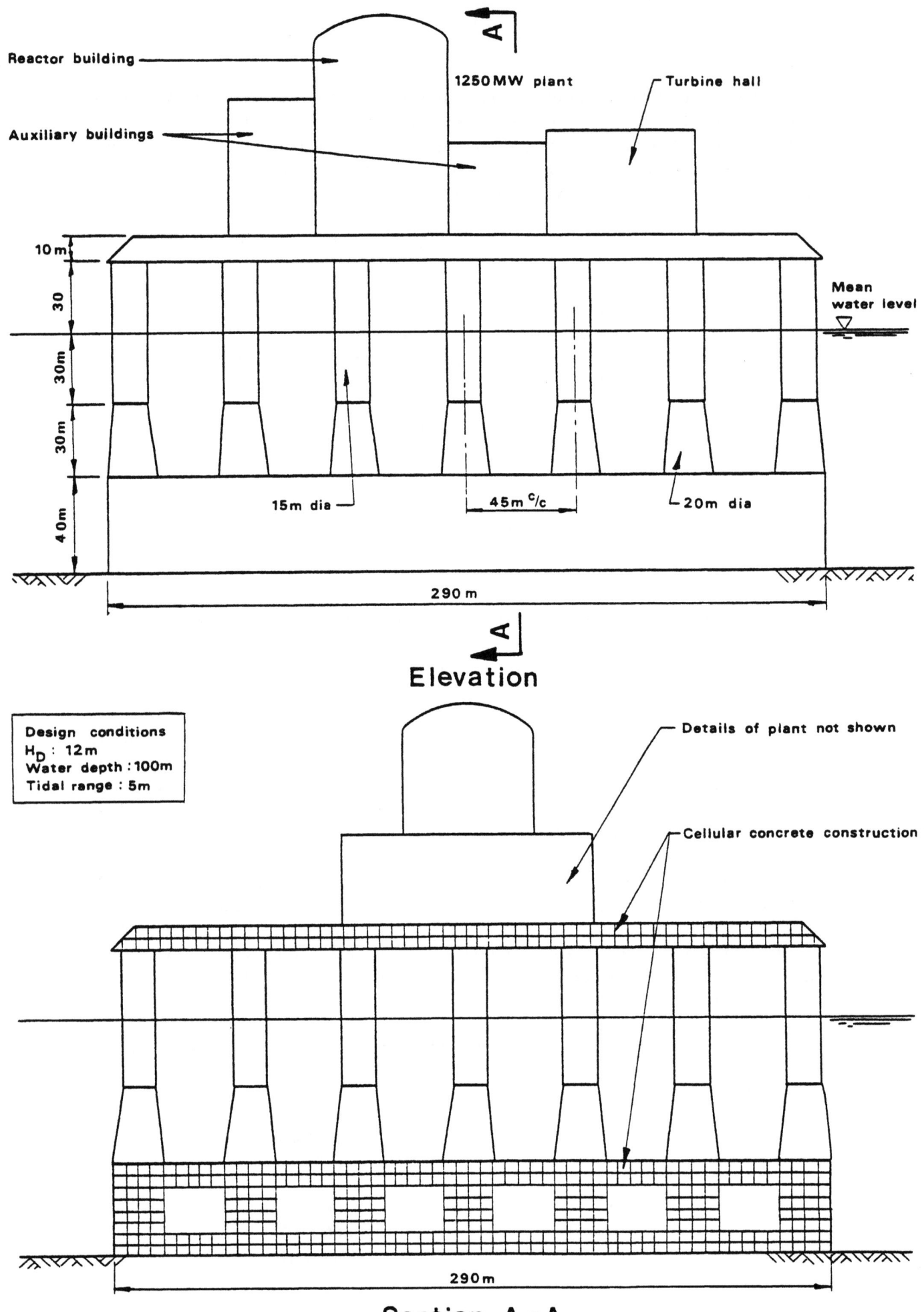

6.45   Unlike a floating island, the vertical forces due to wave action on a semisubmersible are smaller than the horizontal forces. Because of the small water surface area of a semisubmersible, long waves produce little change in uplift forces. For the same reason a tethered semisubmersible is not subject to large changes in buoyancy forces due to tidal rise and fall. Each 1 m rise in water level produces an additional uplift of 85×10^3 kN.

6.46   To limit plant accelerations during operation to 0.02g, mooring forces need to be of the same order as the wave forces. Thus large mooring forces are required. These are not as great as for a floating island and for large waves the reduction is appreciable. When H_D = 12 m the total mooring line tension needed to restrain a semisubmersible is likely to be less than half that required to anchor a floating island.

6.47   With currently available mooring technology several hundred lines would be needed to anchor a semisubmersible. We consider that the technical feasibility of mooring a semisubmersible in exposed conditions is in doubt.

Construction difficulties

6.48   The large size of a semisubmersible island presents considerable construction difficulties. A number of alternative methods of construction have been considered. These alternatives are discussed in detail in Section 7.2.

Summary

6.49   The design concept of a cellular deck supported on cylindrical columns with a cellular buoyancy tank is simple in principle. However, the draught of the island is very large and will cause construction problems. The natural period of the island is greater than the period of high energy waves but the stiffness of the island is low. The mooring system will have to restrain the motion of the island. Although the forces involved are less than for a floating island, they are still large and the feasibility of mooring the semisubmersible island must be in doubt. However, for a deepwater site the semisubmersible appears to have the most potential.

Gravity platform

General

6.50   Structures for water depths up to 100 m and with design waves of heights (H_D) up to 12 m have been considered. The outline designs have been carried out for concrete structures.

Layout

6.51   The plant layout is the same as for floating and semisubmersible islands. A deck area of 6 ha, 245 x 245 m, has been provided.

Design concept

6.52   The concrete gravity platform with fixed legs is similar to the concept for a semisubmersible structure and may be considered as three elements:

>   deck
>   supporting legs
>   base.

6.53   The design considerations for the deck are the same as those for a semisubmersible structure and a similar cellular form of construction has been adopted.

6.54   The supporting legs transfer the deck loads of the structure to the base. Since the legs are subject to wave action it is desirable that they should have a small cross-section near the water surface. However, because stability during float-out depends on the water surface area of the legs they cannot be made too small. Near the base the effect of waves is diminished and a leg of larger section is advantageous to resist bending. A larger leg also helps to spread loads into the base and provide additional buoyancy during float-out. For platforms in water depths up to 100 m we have taken circular legs tapering from 20 m diameter at the base to 15 m diameter in the region of wave action. The legs are arranged on a 45 m grid.

6.55   The base has to distribute the weight of the structure evenly over the seabed and provide buoyancy during construction and tow-out. To give a sufficiently wide leg spacing for roll stability a base of 290 x 290 m has been chosen. This also helps to reduce foundation loadings. A deck of similar area has been provided as this gives support at the top of the legs. However, only the central area of 245 x 245 m is occupied by plant and the remainder is intended to be of lighter construction. The thick base required for buoyancy is also stiff enough to cope with local variations in sea bed conditions. An outline design for 100 m depth of water is shown on Drawing 6.

Draught and freeboard

6.56   The height of the deck soffit above mean sea level has been chosen to provide 2 m freeboard to the maximum wave ($2.1 \times H_D$). The most severe case of a maximum wave occurring in conjunction with a high tide and surge has been considered.

6.57   A gravity platform is not a permanent floating structure and only needs buoyancy during tow-out. The outline designs have not included for the 10% excess buoyancy that was provided for permanently floating structure (see Section 6.11). The structure shown on Drawing 6 is stable when floating with its legs entirely above water or when sunk to the seabed. However at intermediate draughts it is not stable. The structure will thus pass through an unstable state when being sunk to the seabed. To provide for this temporary situation in the design of the structure would add greatly to its size and cost. We have assumed that temporary flotation chambers will be used to maintain stability during ballasting.

6.58 For a tidal range of 5 m and a surge and settlement allowance of 2 m the weights of gravity platforms for 50 m and 100 m depths of water for a range of wave heights are:

H_D (m)	Weight of concrete in island for 50 m water depth (Mt)	Weight of concrete in island for 100 m water depth (Mt)
3	1.71	2.41
6	1.82	2.58
12	1.97	2.71

Construction difficulties

6.59 These structures are larger than the semisubmersible islands and present greater construction problems. Methods of construction are discussed in Section 7.2.

Summary

6.60 The basic design concept is relatively simple. However the draught of the structure when floating is very large and this will cause construction difficulties. Selection of a suitable construction site will be a major problem. The feasibility of gravity platforms for operation at deep water sites is therefore in doubt.

Caisson island

General

6.61 This type of island consists of concrete caissons founded on the seabed with their tops forming a platform above the sea level. It can be formed of several individual caissons, each housing an element of the power station, linked by access bridges. They will require individually prepared foundation pads which need not necessarily be at the same level. The lighter caissons, which require less draught to float, can be positioned around the larger and deeper caisson containing the reactor.

Design concept

6.62 The outline design for the caissons is based on cubical cellular concrete construction. Longitudinal and transverse diaphragms will be required to distribute loads uniformly to the seabed and to span any irregularities in it.

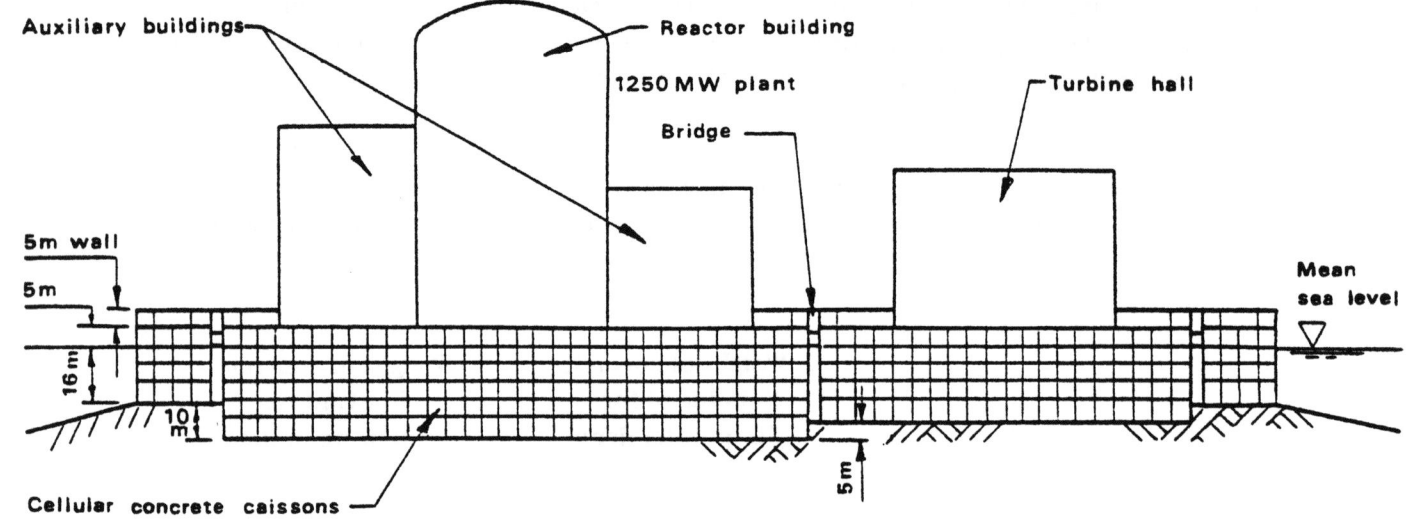

Section A-A

Design conditions
H_D : 3 m
Water depth : 26 m
Tidal range : 5 m

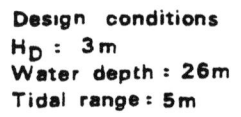

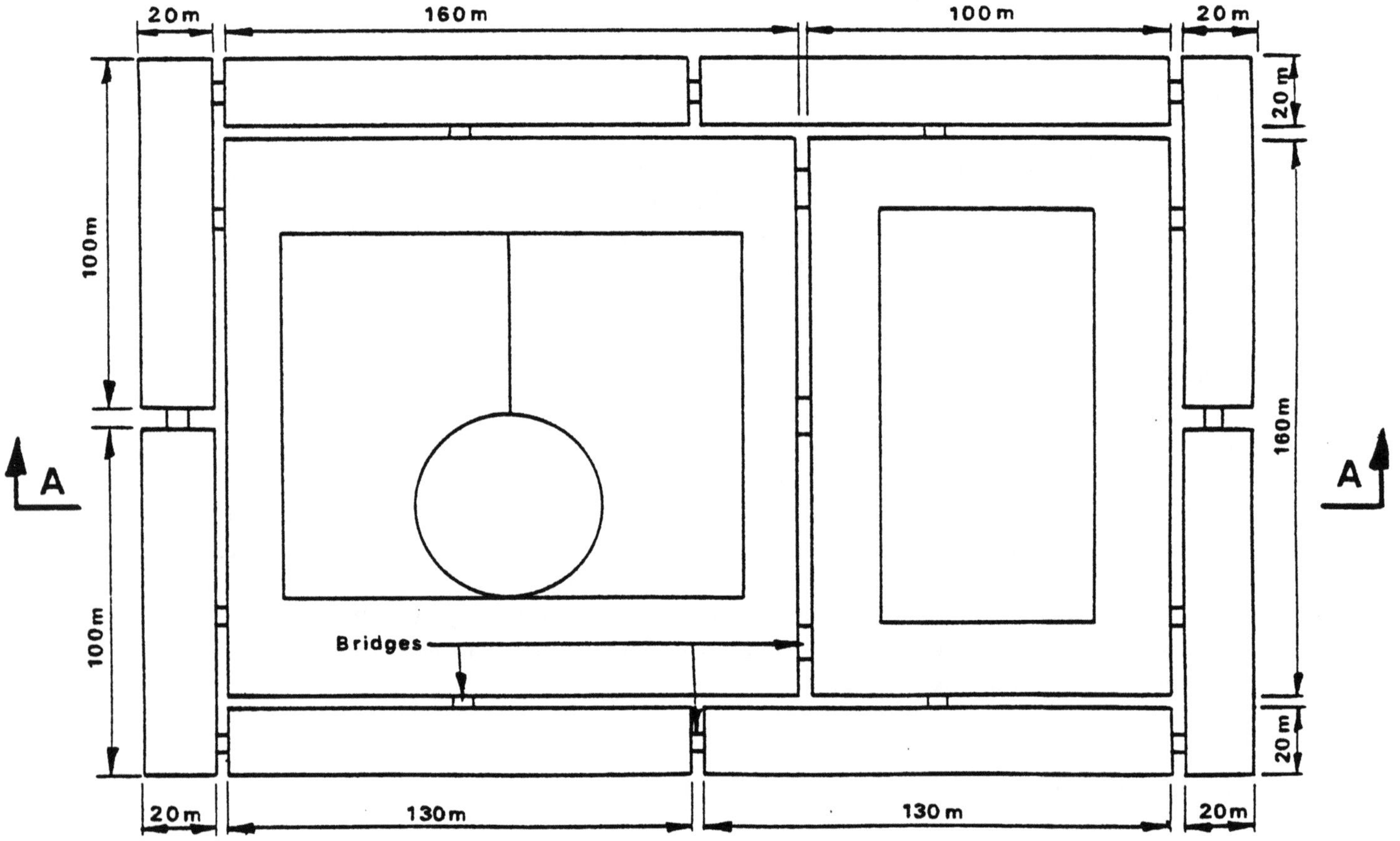

Plan
Scale 1:2000

Caisson island

Drawing 7

6.63   The individual caissons need only be designed to be stable when floated out in reasonable weather conditions. The extra buoyancy required in floating islands for safety will not be required.

6.64   One disadvantage of a caisson island is that it will be best suited to a limited range of water depths. If the water depth of the site is greater than the draught that the caisson requires to float either the caisson must be made higher or fill must be placed on the sea bed to give the caisson the required freeboard when sunk into position. Although this is feasible it will increase the cost of the island. If the water is shallower than the required draught of the caisson when floating, then approach channels will have to be dredged to the site. Extensive excavation will need to be carried out for the foundation pads. Such channels and excavations are feasible but could be expensive; their cost being site specific.

6.65   A typical caisson island would consist of the following individual caissons:

Plant	Caisson size (m)	displacement (Mt)	Min draught (m)
reactor	160 x 160	0.61	25
turbine hall	100 x 160	0.26	20
Edge caissons	4 No 100 x 20	0.01	15
Edge caissons	4 No 130 x 20	0.015	15

Fill island

General

6.66   The major variables considered in the designs are:

(a) platform area
(b) crest level
(c) platform level
(d) width of crest and backslope
(e) slope protection
(f) shape of island.

6.67   The choice of values for these variables is largely determined by the plant requirements and site conditions. The way that island variables have been considered in relation to plant and site constraints is described below.

6.68   For given plant and site requirements it is possible to design several technically satisfactory islands. The difference between the islands lies in the form of sea protection adopted. The form of sea protection is discussed further in Sections 6.74 - 6.79.

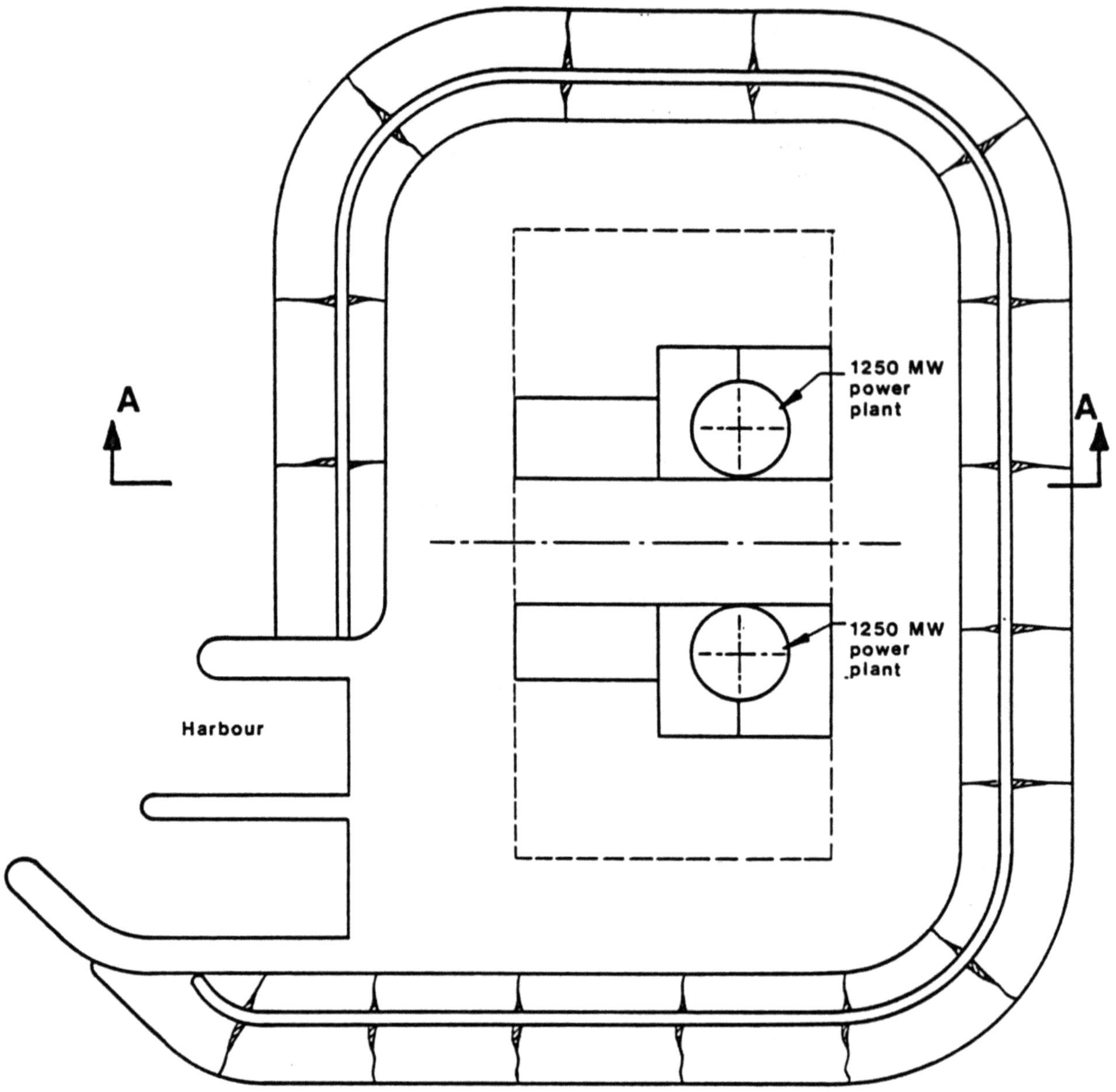

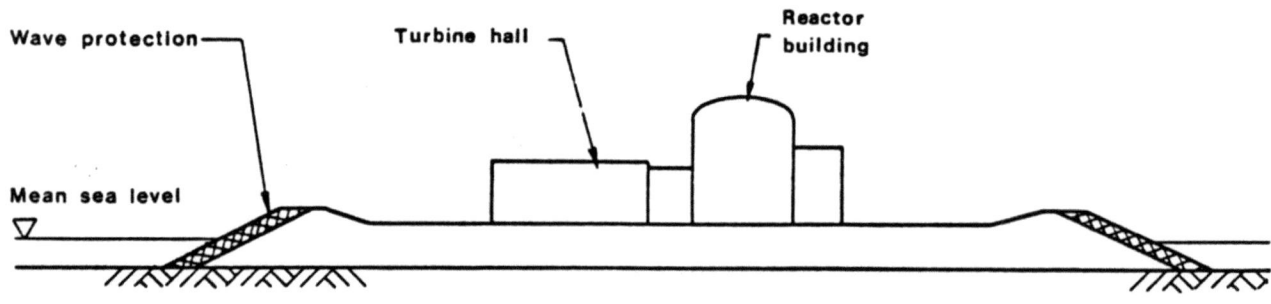

Platform area

6.69   The terms of reference require provision of a 6 ha site for a 1,250 MW plant and consideration of a total area of 15 hectares for the combined plant and auxiliaries. Thus the range of island size is from 12 ha to 120 ha for 2,500 MW to 10,000 MW developments.

6.70   A possible layout for a 2,500 MW development is shown on Drawing 8. The total platform area is 30 ha, the main plant occupying 12 ha. A 30 ha island has been illustrated as the total cost of the project, including the power plant, will generally be less for a 30 ha island than for a 12 ha island. This is because the additional area provides an on-site construction area for the main plant which would otherwise have to be constructed using a shore base and minimum on-site area. The saving on the power plant construction is likely to be greater than the increased cost of the island, for most locations. The additional area provides a zone of 100 m minimum width between the main plant and the edge of the island. This zone can be eventually used for access, auxiliary structures and services. It will also help to minimise the effects of spray on the main plant.

Crest Level

6.71   Crest levels have been selected so as to ensure that overtopping will not occur during the design storm. The height above mean water level to satisfy this condition depends on the wave run up which is a function of the wave height, type of slope protection and gradient of the slope. The table below shows the maximum run-up for various slopes during a storm with a 5 m design wave height. The value given for the run-up is the maximum height above the still water level:

Design wave height (H_D)	Max wave height (H_{max})	Slope type	Gradient	Max vertical run-up (R_{max})
m	m			m
5	10.5	Smooth	1 on 3	24.2
5	10.5	Smooth	1 on 6	10.7
5	10.5	* Rip rap	1 on 2	10.0
5	10.5	Rip rap	1 on 6	4.6
5	10.5	* Dolosse	1 on 2	10.0

* *Rip rap is the name given to large rock produced within a particular range of sizes. Dolosse are a type of interlocking concrete unit – see drawing 10.*

NOTATION

Design wave height	H_D
Maximum wave height	H_{max} ($:2.1 \times H_s$)
Tidal range	T
Allowance for settlement and surge	S
Maximum wave run-up (Corresponding to incident wave of height H_{max})	R_{max}

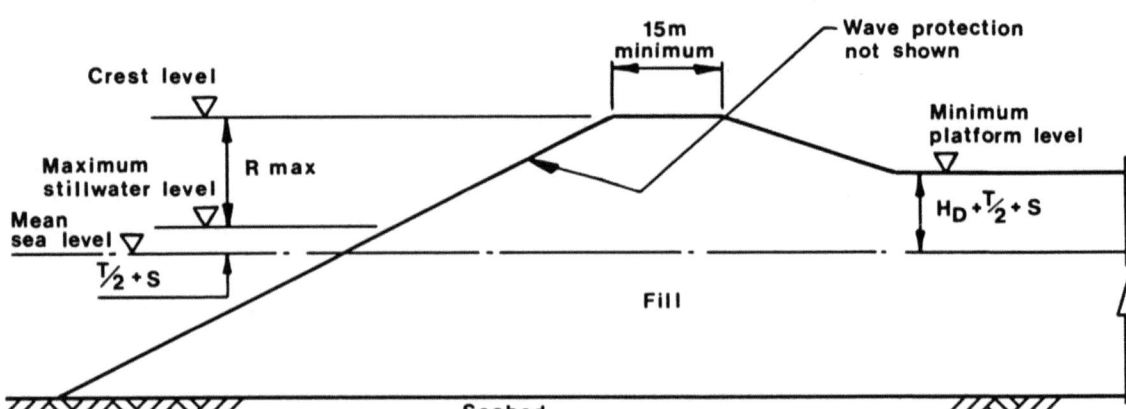

Fill island – typical slope detail

Drawing 9

Dolos armour units

Drawing 10

6.72   In the design the conservative assumption is made that the most severe wave conditions are combined with the maximum still water level. The maximum still water level is taken as the mean sea level plus half the maximum tidal range plus an allowance for surge and for settlement after construction. Surge will vary considerably between sites, for example being much greater in the southern North Sea than in the Mediterranean. Extreme water levels are discussed in Section 4.50 and Appendix J. A typical slope profile is shown on Drawing 9.

Platform level

6.73   This is the general ground level of the island. Some saving on the volume of fill material can be made if this level is made lower than the crest. However, if the platform level is made too low the risk of flooding and the problems of drainage associated with polder-type islands are encountered. In the designs the minimum platform level has been taken as mean sea level plus the design wave height added to half the maximum tidal range and the surge and settlement allowance.

Width of crest and backslope

6.74   The minimum crest width is limited by the need to maintain sufficient space on the top of the bank for access to the seaward slope during construction and for maintenance. A minimum crest width of 15 m provides sufficient space for a large crane with outriggers to place sea protection units. The steepness of the backslope is limited by stability considerations and a gradient of 1 on 3 has been taken for this slope.

Sea protection

6.75   Protecting the island against wave attack is a major part of the cost of fill islands. The effects on cost are both direct: the cost of the slope protection materials; and indirect: the influence that slope protection has on the crest level and gradient of the seaward slope and hence the volume of fill. A number of different types of protection have been considered and a comparison of the different types is summarised in Table 4. A serious limitation on some protection systems is the difficulty of constructing them below water level. For this reason we have concentrated our studies on:

(a) large graded rock randomly placed to a specified thickness, referred to as rip rap

(b) precast concrete blocks, randomly placed to a specified thickness; many different shapes of block are potentially suitable including dolos and tetrapod units.

(c) artifical beaches of sand or gravel.

6.76   Typical profiles of slopes protected by the first two methods are shown on Drawings 11 and 12.

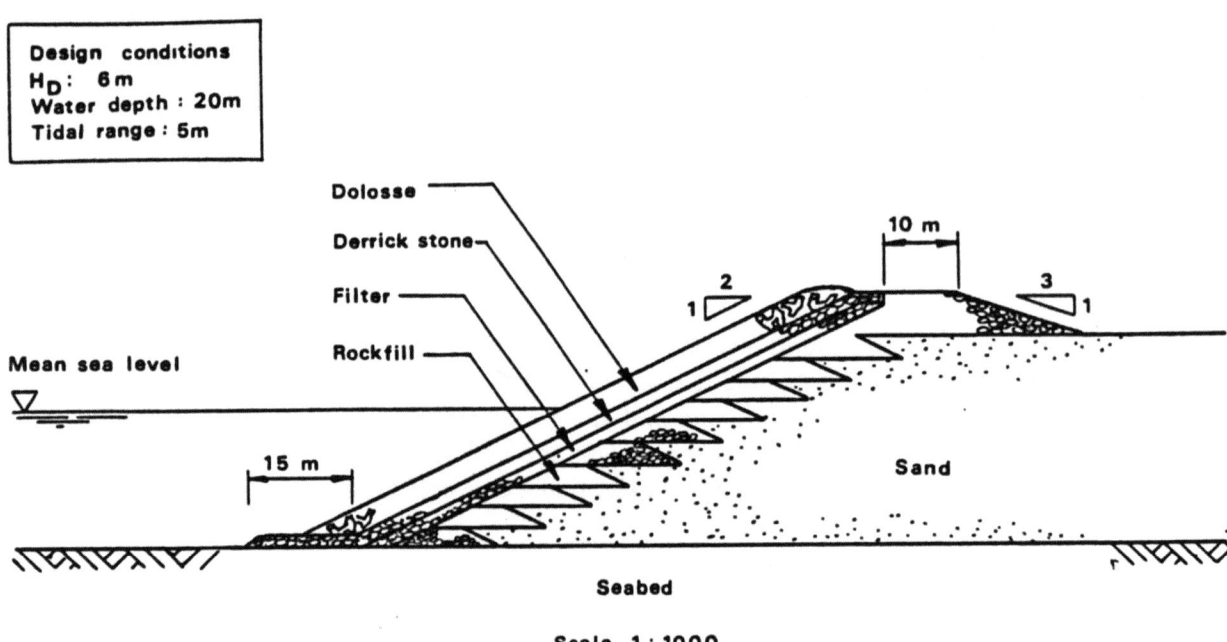

Wave protection - Dolosse Drawing 11

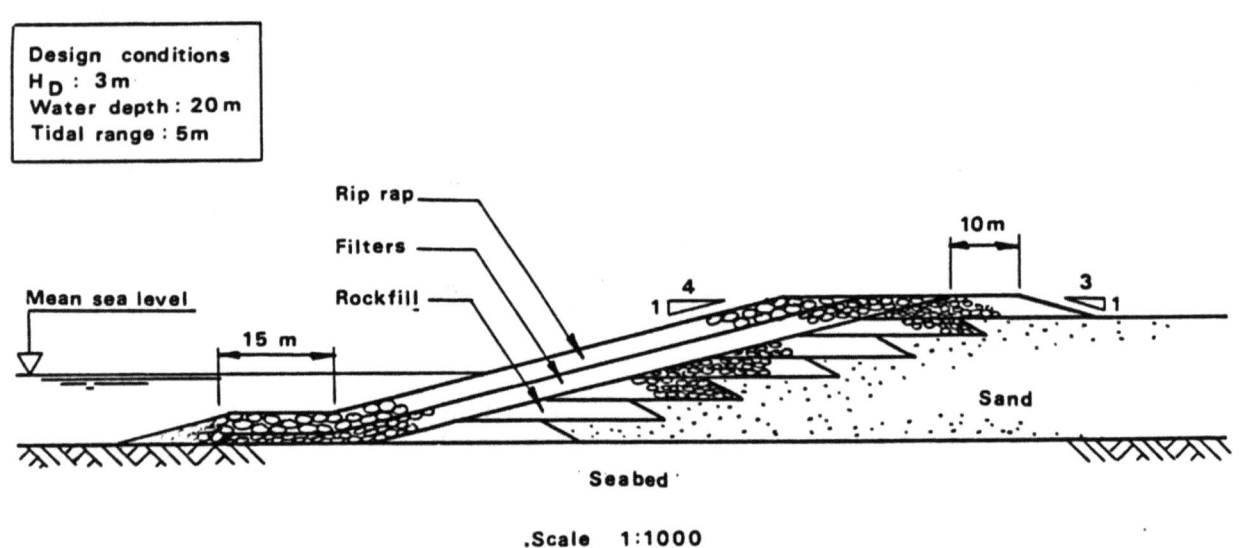

Wave protection - Rip rap Drawing 12

TABLE 4   COMPARISON OF ALTERNATIVE SEA PROTECTION SYSTEMS

	Artificial beach	Rip rap	Concrete blocks regular pattern	Concrete blocks random orientation	Rip rap with bitumen grout	Concrete blocks with bitumen grout	Stone filled mattresses	Concrete slabbing	Asphaltic concrete slabbing or mattress
Design & Construction									
Design rules developed	part	yes	yes	yes	part	part	part	part	yes
Transport by sea possible	yes	yes	yes	yes	yes	yes	yes	no	yes
Can be placed under water	yes	yes	no	yes	yes	no	yes	no	yes
Progress in poor weather	some	yes	no	yes	some	no	some	no	some
Performance									
Experience in exposed marine conditions	some	yes	yes	yes	yes	yes	some	no	yes
Damage easily monitored above low water	no	no	yes	yes	yes	yes	yes	yes	yes
Self-healing properties	partial	yes	no	yes	no	no	no	no	no
Repairable in poor weather	yes	yes	no	yes	no	no	no	no	no
Resists accidental damage (eg oil spillage)	yes	yes	yes	yes	part	part	no	some	no

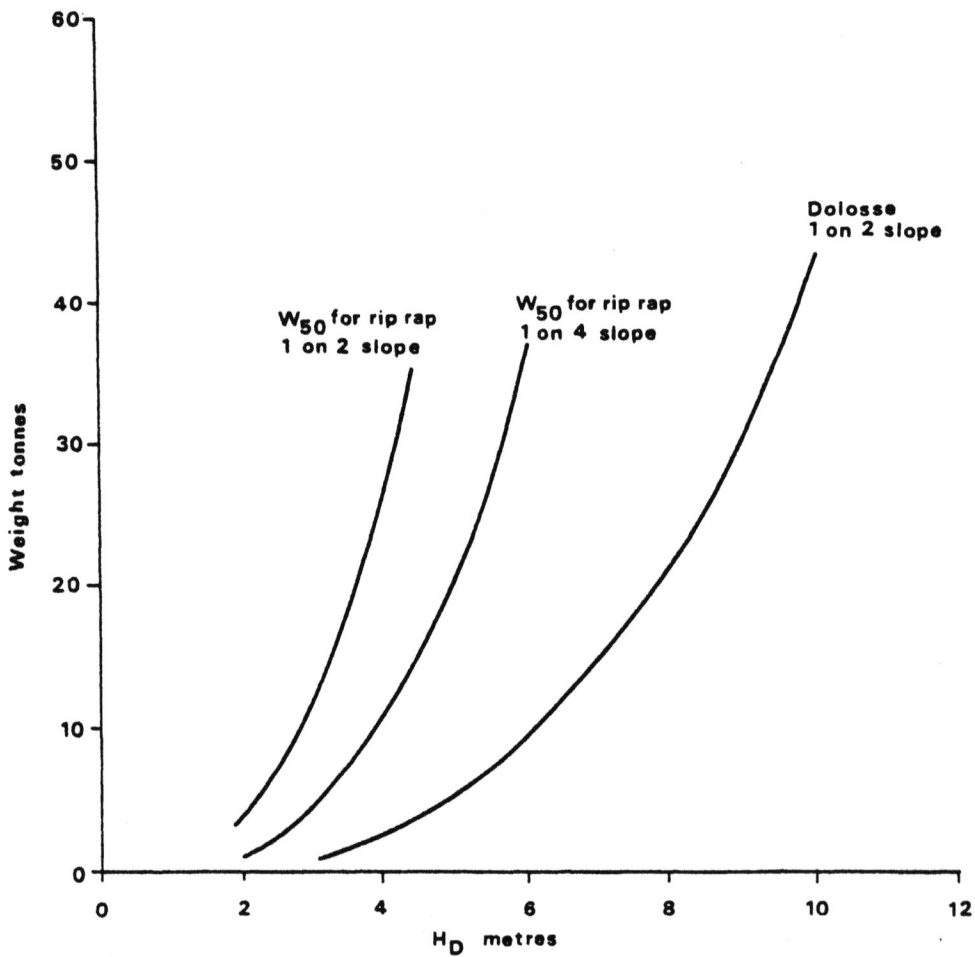

Dolosse and rip rap protection
Weight/wave height relationships

Drawing 13

6.77 Dolos concrete armour units have been chosen for our outline design with interlocking concrete block protection. A breakwater constructed from dolosse is shown on Drawing 10. Dolosse have good interlocking properties so that comparatively small blocks can resist large waves. Many other shapes of armour unit have been devised and several, including tetrapods, have stabilities approaching those of dolosse. Drawing 13 shows the weights of dolosse needed to resist waves of varying heights.

6.78 Rip-rap is an alternative to concrete armour blocks. Because rip-rap does not interlock in the same way as dolosse, much larger individual pieces are needed. Rip-rap is a mixture of pieces of rock of varying size. A convenient measure of the grading of the rock is the D_{50} size. This is the length of the side of a square opening through which 50% by weight of the material will pass. If the rock density is known, a median weight of rock, W_{50}, may be derived from the D_{50} size. Median weights of rock for varying wave conditions and slopes of 1 on 2 and 1 on 4 are shown on Drawing 13. The largest rocks will weigh much more than the median weight, the ratio W_{max}/W_{50} depending on the grading. As can be seen from Drawing 13 the median rock size increases rapidly with increasing wave height. For design wave heights above 3 m the sizes of rock required present practical difficulties of production and transport.

6.79 There is less experience of the use of artificial beaches of sand or gravel than of concrete block or rip-rap protection. Design rules for beaches are not fully established. The stability of a beach is largely determined by the grain size of the material from which it is formed. Site specific conditions such as the strengths and directions of currents may have a major influence on the rate at which material is eroded from a beach. Maintenance can form a large part of the cost of this type of sea protection. A typical slope for a sand beach with a median grain size of 0.25 mm would be 1 on 50.

6.80 The costs of the three different types of slope protection are compared in Section 10.

Shape of island

6.81 The following factors influence the shape of the island:

> layout of the plant
> exposure and cost of sea protection
> cooling water dispersion
> future extensions
> site specific problems.

6.82 The optimum layout of the plant appears to be in pairs back to back with successive pairs in line. This simplifies power transmission, access and services and also allows cooling water to flow across the island. However, it results in a high length to breadth ratio for the final 10,000 MW development.

6.83 The sea protection is a major part of the cost of the island. Minimizing the island's exposure to any prevailing storm direction can reduce costs. Similarly minimizing the overall perimeter is desirable. A 10,000 MW development which minimizes the exposed perimeter of the island is shown on Drawing 21.

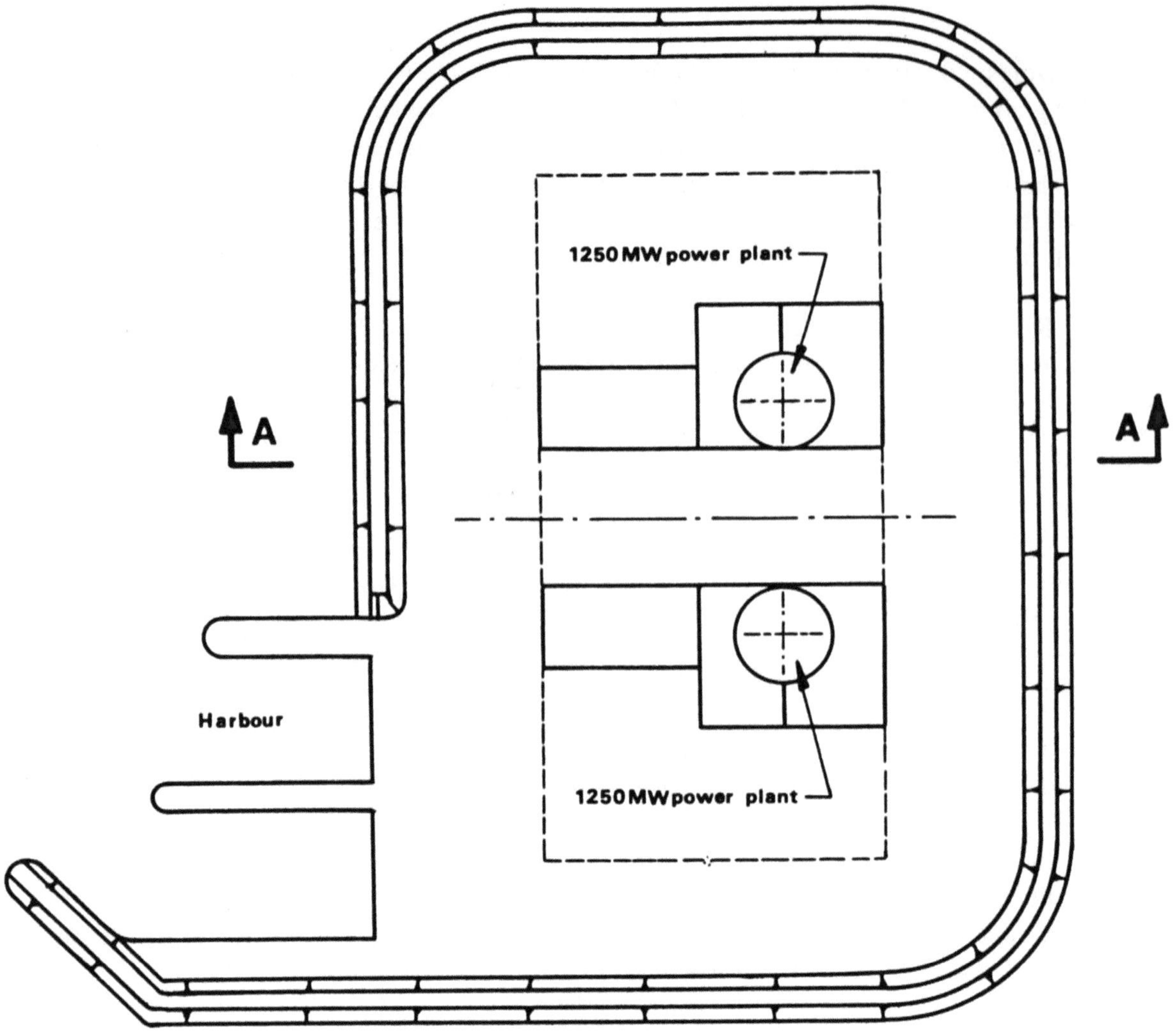

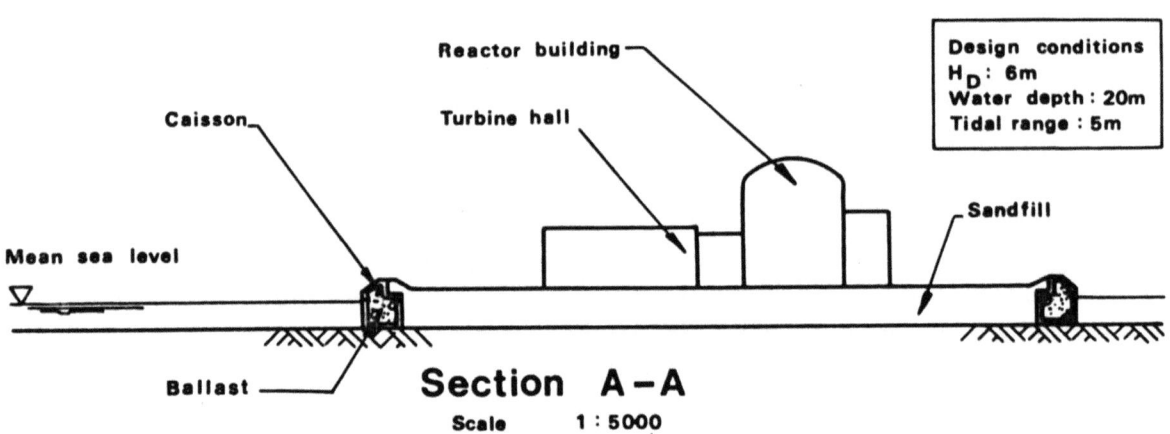

Plan
Scale 1:5000

Design conditions
H_D: 6m
Water depth: 20m
Tidal range: 5m

Section A-A
Scale 1:5000

Caisson retained island Drawing 14

6.84   For optimum cooling water dispersion a high length to breadth ratio will probably offer the best solution.

6.85   The final shape of the island may well be determined by combinations of the above and site specific problems such as sea bed topography, foundation conditions, scour, adjacent coastline erosion or accretion and the effect on shipping.

Summary

6.86   The fill island is a technically simple concept which has been developed both in theoretical studies and by actual construction. The additional space required for working areas can be used for auxiliary buildings at no extra cost.

Caisson retained island

General

6.87   The caisson retained island is in effect a fill island with sea protection provided by caissons. The general comments on fill islands apply equally to the caisson retained island. This section illustrates the caisson design only. A 30 ha island is illustrated, in Drawing 14, for the reasons given in Section 6.70.

Design concept

6.88   The caissons can be of cubical cellular or cylindrical cellular concrete construction. They will be designed to be self buoyant and constructed in a dry basin. The foundation will be levelled before the caissons are towed out and sunk into position. The caissons will be filled with sand to increase their stability. Later caisson units might be fabricated on the island and placed by crane. The toe of the caissons will probably require some form of scour protection, the extent and nature of which will depend on the water depth, design wave height and tidal currents. Rip-rap, concrete or asphalt mattresses could be used for the scour protection.

6.89   Sizes of caissons for various water depths and design waves are shown below. A tidal range of 5 m has been assumed. The lengths of the caissons will be about 1.5 times their widths.

		Water depth (m)		
		10	20	30
$H_s = 3$ m	Height (m)	20.5	30.5	40.5
	Width (m)	16.0	27.0	36.0
$H_s = 6$ m	Height (m)	26.5	36.5	46.5
	Width (m)	20.0	30.0	41.0

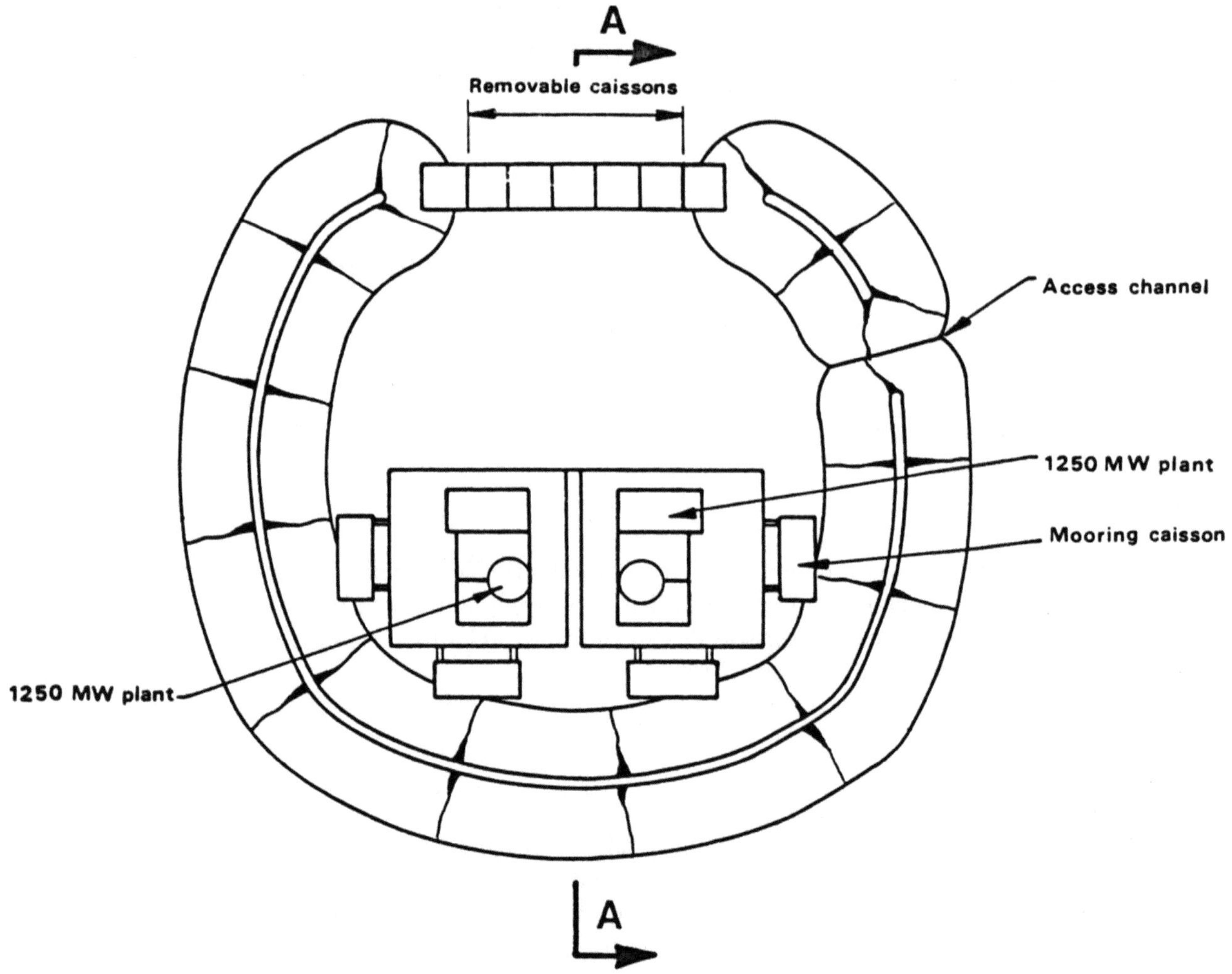

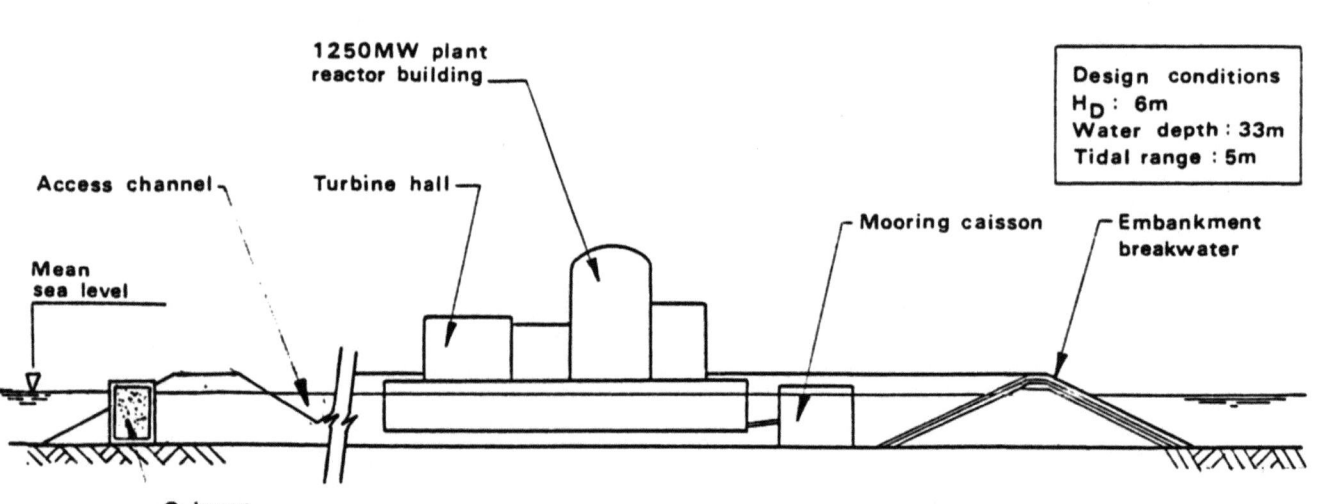

Protected floating island

General

6.90   The protected floating island is a floating island which is moored within a breakwater. The breakwater protects the island from the effects of waves and thus eliminates the need for pre-loaded moorings which are used to limit the movement of floating islands. It also shelters the island from collision. The reasons for doubting the feasibility of the floating island do not apply to the protected floating island. A 12 ha 2500 MW development is illustrated in Drawing 15.

Design concept

6.91   The raft within the breakwater will be similar to that described under floating islands except that the freeboard against wave attack can be reduced and thus the overall weight and draught will be less. Assuming a design wave (H_D) of 3 m within the breakwater the principle statistics of the raft are:

Plan dimensions	245 x 245 m
Weight of raft	0.59 Mt
Total weight	1.20 Mt
Draught	20 m

6.92   The mooring system must be capable of holding the raft in position without excessive horizontal movement because this would require adjacent rafts to be moored further apart and thus increase the length of the breakwater required to enclose them.

6.93   The breakwater will require the same sea protection as a fill island and the same design concepts apply. One section of the breakwater will be formed from caissons to allow the float in and removal (if required) of the floating island.

6.94   Pneumatic breakwaters have been considered but they are too costly to operate and are not effective for the longer period waves. The pneumatic breakwater allows air to bubble up from the sea bed producing surface currents which cause the incoming waves to break. Floating breakwaters consisting of moored rafts have not been developed sufficiently to provide a viable protection system.

6.95   A mooring system employing articulated struts anchored into caissons incorporated into the back of the breakwater or free standing would be best suited for sites with an appreciable tidal range. Preloaded cable moorings could be used if the tidal range is small.

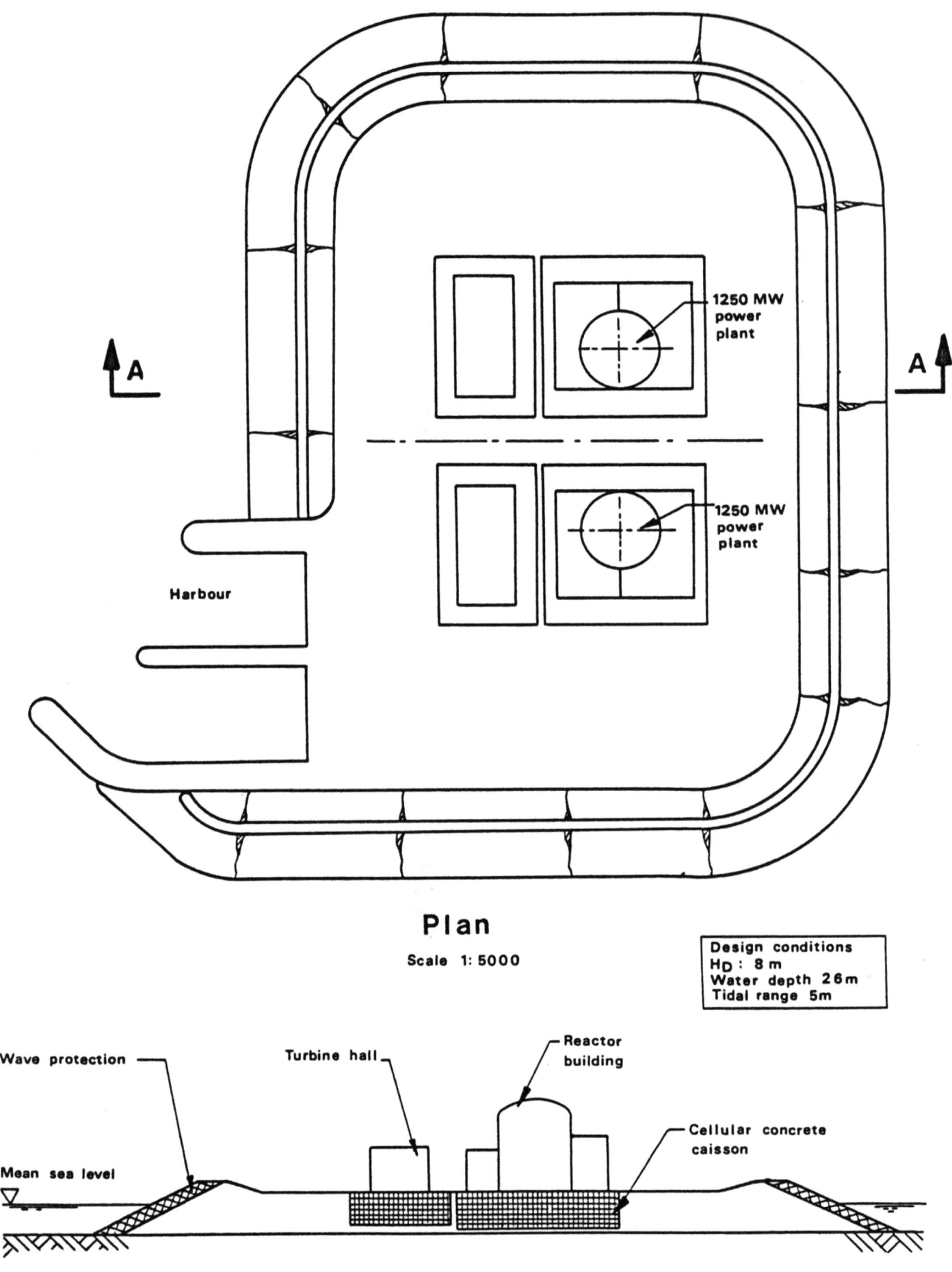

Caisson fill island

General

6.96 The caisson/fill island is essentially a fill island which is constructed so that some of the main plant elements can be floated in on caissons before the fill is completed. This could result in saving of time, and therefore money, as it will be not be necessary to wait for the island to be partially completed before starting the power station construction. However, the time to construct the caissons themselves will be considerable and time savings relative to an ordinary fill island are unlikely to be achieved unless existing facilities can be used for constructing the caissons. A 30 ha island is illustrated, in Drawing 16, for the reasons given in Section 6.70.

Design concept

6.97 The caissons will be similar to those for the caisson island except that as they will be protected from the sea by the surrounding sand fill, they will not require the extra strength and height required for the caissons exposed to wave attack.

6.98 The fill part of the island will be identical to the ordinary fill island.

Caisson/piled island

General

6.99 The caisson/piled island has the same basic philosophy as the caisson/fill island. The construction of the power plant on caissons can be carried out independently of construction on site. The caissons will be designed to withstand wave attack as the piled surround will not provide much shelter. A 12 ha island is illustrated in Drawing 17.

Design concept

6.100 The most suitable types of pile for use in a marine environment are driven piles. Bored piles are less suitable for underwater construction. The two basic types of driven pile are the large displacement type - steel box and precast concrete piles - and the small displacement type - open ended box and steel H piles.

6.101 Box and tubular steel piles are usually used for marine structures. Precast concrete piles are heavy and difficult to handle if long lengths are required. Steel H piles present larger areas for corrosion and have greater drag coefficients when waves and currents are considered.

Section A - A

Design conditions
H_D : 3m
Water depth : 26m
Tidal range : 5m

Plan
Scale 1: 2000

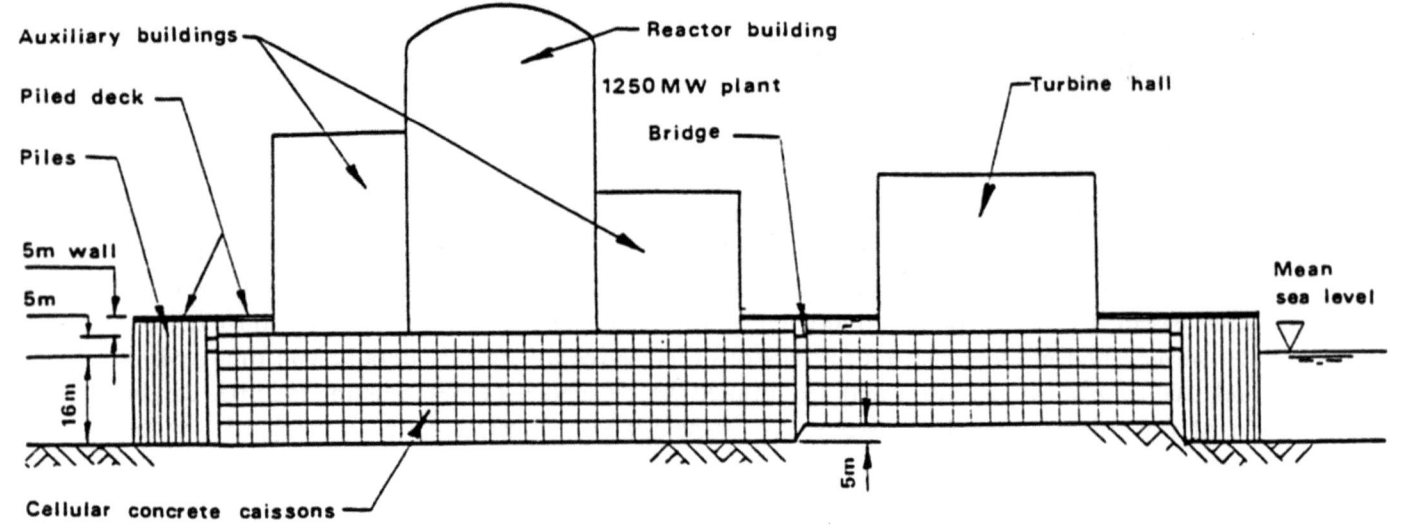

Caisson/ pile island

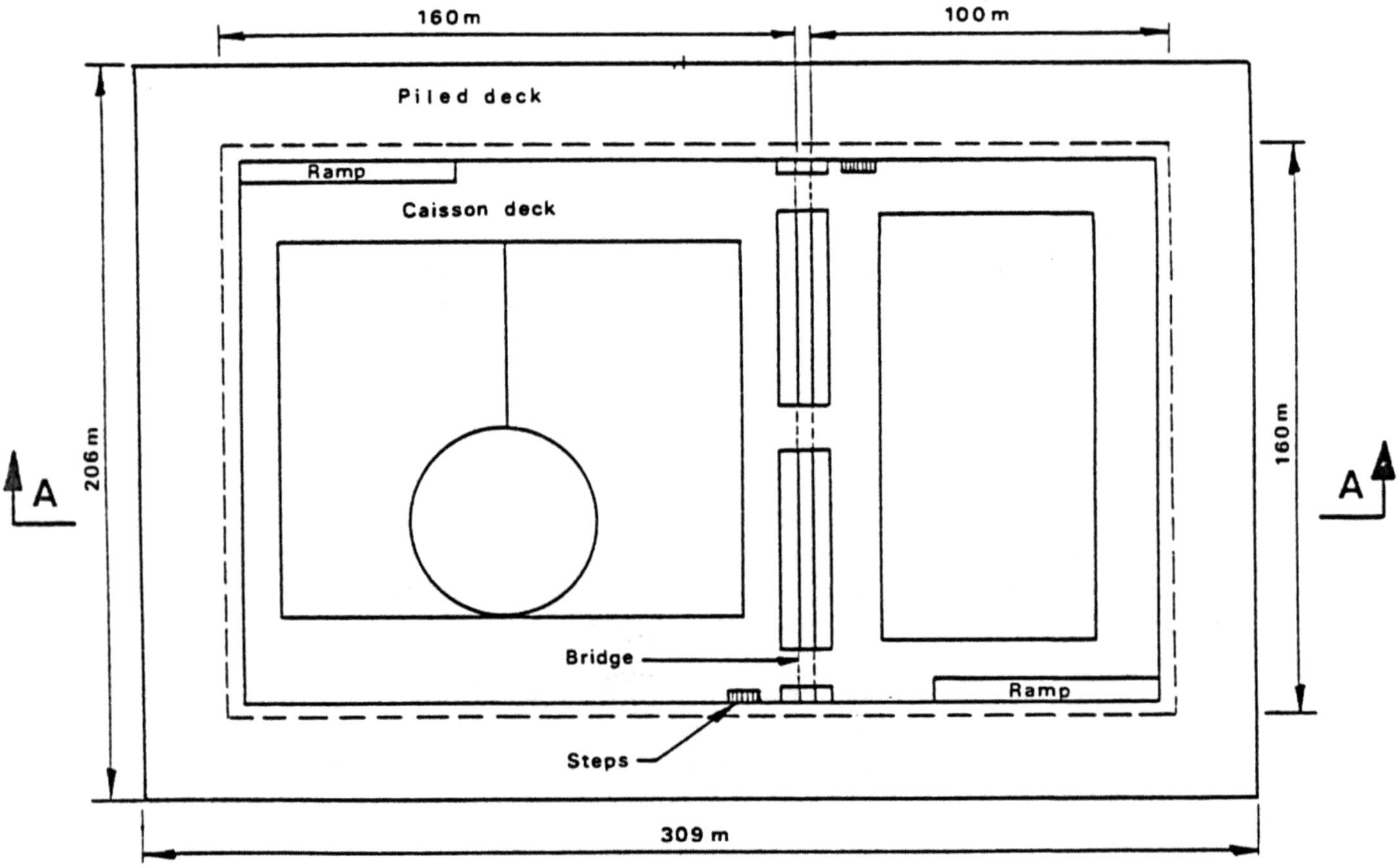

Drawing 17

6.102   Steel piles will probably require cathodic protection. Protection of the splash zone for a design life of 80 years poses a major problem. For this reason we have based our design on precast concrete piles despite the difficulties of handling them in long lengths.

6.103   The pile size, and penetration will depend on the properties of the sea bed at the specific site. If the seabed is medium dense sand then a piling pattern using 900 mm diameter piles on a 4 to 5 m grid with 10 to 15 m penetration would be adequate to support the auxiliary buildings and equipment.

Cooling water dispersion

6.104   For efficient operation of the cooling water system both primary and secondary recirculation must be minimized. Primary recirculation is the direct transfer of heated water from the outfall to the intake. Secondary recirculation occurs as a result of an overall background temperature rise in the surrounding water.

6.105   The risk of primary recirculation can be minimized by vertical and horizontal separation of the intake and outfall and, in general, techniques developed for conventionally sited power stations should apply.

6.106   Secondary recirculation is likely to occur to some degree at all sites. The background temperature rise depends on many factors of which water depth, tidal currents and dispersion characteristics are the most important. It is only possible to predict results for a specific site and extensive data have to be acquired. Secondary recirculation can be reduced by careful siting of the intake and outfall and the use of diffusers on the outfall.

6.107   A general technical note on the dispersion of cooling water is given in Appendix L. The effects of water depth, residual currents and tides on the background tempeature is examined. Using ranges of typical values the possible conditions in the North Sea and the Mediterranean are indicated.

Access

Types of access

6.108   The design criteria for access are considered in a broad way in Section 4.67.

6.109   For those types of island that are constructed in wet and dry basins the access problems during construction are less severe as land routes are available. For islands that are constructed offshore and for the operation of all islands the access requirements can be met either by sea or air transport or a fixed access. Sea or air transport will have to be used until the fixed access is established.

6.110 Sea transport could be by one or combinations of the following

>conventional ship
>ro-ro ship (roll on - roll off)
>hydrofoil
>hovercraft

6.111 Helicopters could be a viable alternative to hydrofoils or hovercraft for transporting personnel to remote locations. They are extensively used in the North Sea oil field. A helicopter landing pad will be required at all islands.

6.112 A fixed access could be by one or a combination of

>road embankment
>bridge
>bored tunnel
>submerged tube tunnel

Floating islands, semisubmersible and gravity platforms

6.113 For these types of island a fixed access will not be practicable because of the siting of the island. Sea transport will therefore have to be used. The provision of an all weather harbour will only be possible if the island is sited in water depths at which a breakwater type harbour is feasible. For these types of island water depths will generally be too great to construct breakwaters. Consequently access to these islands will be restricted.

Other types of islands

6.114 For the other types of islands a harbour can be incorporated in the design of the island. A work harbour will in many cases be required during their construction and it can be used as a permanent harbour by ensuring that the design is adequate for the design life of the island. Alternatively the design should allow for the replacement of major elements during maintenance without serious interruption to traffic.

6.115 The required size of the work harbour will be larger than that for the permanent harbour.

Fixed access

6.116 A road embankment will be less vulnerable than a bridge to damage by collision or other accident. It will, however, block navigation. It could cause sedimentation problems but in some conditions could be used to reduce the recirculation of cooling water. A bridge will not significantly influence either of these aspects. Tunnels have the advantage of not being liable to collision damage but they are still vulnerable to accidents inside them.

6.117   A road embankment will be formed by the method used for fill islands - a 12 m carriageway has been assumed. A concrete box girder bridge with 30 m spans supporting a 12 m carriageway has been taken for comparison. Elevated navigation spans have not been allowed for. The tunnel solution has been based on two 10 m diameter tunnels both containing a 7.5 m wide roadway.

6.118   If a fixed access is to be relied upon at all times then the vulnerability of a bridge may require it to be duplicated. If not, alternative sea transport should be readily available in case of emergency. We have taken two smaller tunnels rather than one large one because of increased reliability.

6.119   A fixed access has several major advantages over sea transport namely:

> speedier access
> less rehandling of materials
> less disruption in bad weather.

These advantages will have a large effect on the costs of island and plant construction.

Work harbour

6.120   The work harbour must be established as early as possible in the construction programme. Sheet piled berths are therefore preferable as they can be completed earlier. However, their maximum design life is shorter than a diaphragm wall or concrete block work quay. In exposed waters caissons may be attractive. Their construction time will be greater than for the other solutions. However, provided suitable shore facilities are available, caisson construction can start before any work on the island site. Thus the completion date for a caisson harbour could be similar to the other solutions. For costing purposes we have taken a sheet pile quay with tie backs to pairs of raking piles, and allowed for replacement during the design life of the island.

6.121   The breakwaters protecting the harbour will be an extension of the sea protection around the island.

6.122   The work harbour will not only have to provide berthing facilities but also be large enough to shelter the less sea worthy marine plant such as work boats and barges in the event of bad weather. Its size will depend on the type of island and the extent of the development. The minimum size is that required during operation and is indicated below:

Power station development	No. of vessels				Total quay length
	Ro-ro	Ferry	or	Hydrofoil	
2500 MW	1	1	or	2	240
7000 MW	1	1	or	3	300
10,000 MW	1	2	or	6	450

The ro-ro vessel is 90 metres long. The 60 m long ferries carry 500 people and the 35 m long hydrofoils 120 people.

6.123   If hovercraft are used instead of ferries then an approach ramp some 50 m wide will be required with parking facilities for the hovercraft.

7. CONSTRUCTION METHODS AND TIMES

Introduction

7.1 The discussion of outline designs has shown that although several are technically adequate there is doubt over the feasibility of constructing some of the structures. In this section methods of construction are discussed and alternative ways of building floating islands, semi-submersibles and gravity platforms are compared. The development of the basic 2500 MW station to 5000 MW and 10,000 MW is considered in Section 7.63 onwards.

Floating and semi-submersible islands

General

7.2 Three possible methods by which these types of island could be built have been considered. In outline these methods are:

- A. Dry and wet basin construction. Most of flotation raft is built in a dry basin which is then flooded allowing the raft to be towed to a wet basin. At this deeper water site the raft is completed, the remainder of the island is constructed and the nuclear plant is installed. The island with its plant is then towed out to its operating site.

- B. Single basin construction. Complete construction of the island and installation of plant is carried out in a dry basin. This basin is flooded to allow the finished island to be towed out to its operating site.

- C. Modular construction. The raft of the island is divided into a number of smaller modules. Each module is built in a dry basin and then towed to the wet basin where the raft sections are joined to one another. Construction then proceeds as for A above.

7.3 One other approach to building a semi-submersible island has also been investigated. This is:

- D. Double dry basin construction. The flotation raft with legs is built in one dry basin. Simultaneously the deck and the nuclear plant are built in a second dry basin. The raft and legs are towed out to a sheltered deepwater site where they are ballasted down. Using the deck for temporary buoyancy the plant and deck is towed over the legs and connected to them. The completed structure is then towed to its operating site.

7.4 These four methods, their site requirements and outline construction are discussed in the following sections.

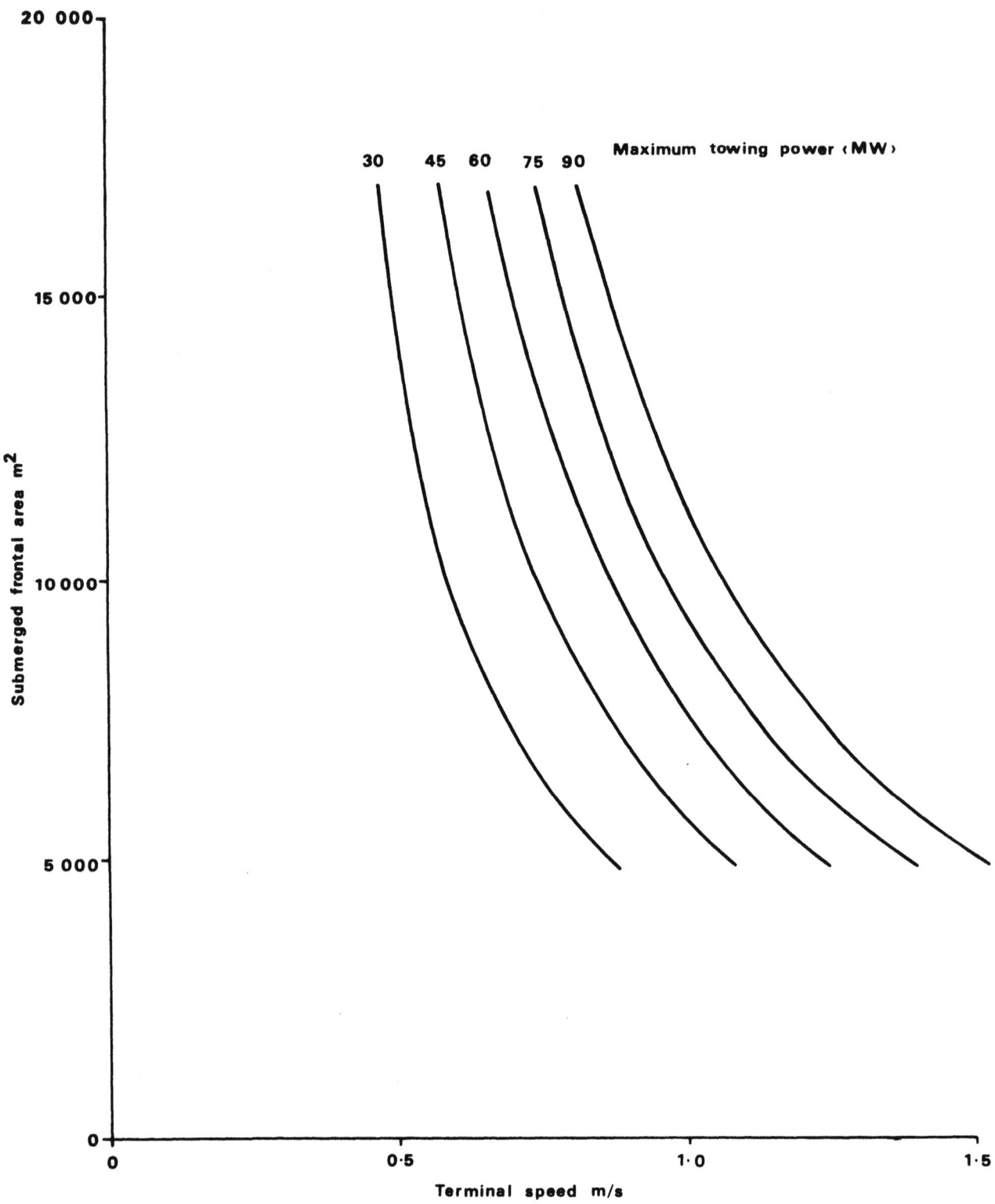

Towing power/frontal area/speed relationship

Drawing 18

Dry and wet basin construction

7.5   This method requires the use of two basins for the main island. Auxiliary islands can either be built in the same basin concurrently with the main island or in smaller separate basins.

7.6   The initial basin is a dry basin which will be of the order of 8 m deep when flooded. The basin will be sited on the foreshore or reclaimed land. The excavation will be kept dry by a wellpoint dewatering system. On the floor of this basin the base slab of the island will be cast and the walls and diaphragms started. Overhead cableways will supplement tower cranes for placing concrete and moving materials. When the walls and diaphragms have reached about 20 m height, the structure will be structurally able to resist the forces imposed on it by flotation. The degree to which raft construction can be completed in the dry basin will depend on the depth of water available. The float out of the partially completed island will be achieved by flooding the basin and winching it out along a previously dredged channel into deeper water. It will then be towed to the wet basin.

7.7   The wet basin is the fitting out basin. Its main requirement is that it should have sufficient depth available for the island to be structurally completed and the power plant installed. The depth required for this varies according to the type of island but is not less than 30 m. Sites at which such depths can be found relatively close to the shore are few. The basin must give access to the full length of one side of the island. To achieve this a row of caissons 300 m long will be required to form a vertical face against which the island can be moored. These caissons can be constructed in the same dry basin as the island to save cost. A rockfill platform behind the caissons will form the working area. This will be linked to the shore by a rockfill embankment to provide access. The cost of the wet basin will be high. The cost of any increase in the working area will have to be judged against the savings from reduced congestion at the site. The walls, diaphragms and floors of the island will be completed once it is moored in position at the wet basin. The island will be ballasted to keep the working level at approximately the same level as the working platform. Once the structure of the island has been completed, the construction of the power plant on the island can be started. The island will be deballasted during this period to keep it level with the working platform.

Tow-out

7.8   On completion of the power plant installation on the island, it is ready to be towed to its final location.

7.9   The towing of vessels of the size of a floating island or semisubmersible has not been attempted before. The maximum speed that can be achieved is small. Although a suitable weather window, during which wind and waves are small, can probably be achieved, the adverse effects of high tidal currents cannot be overcome. This will impose restrictions on the site for the wet basin as only areas without large tidal currents will be suitable.

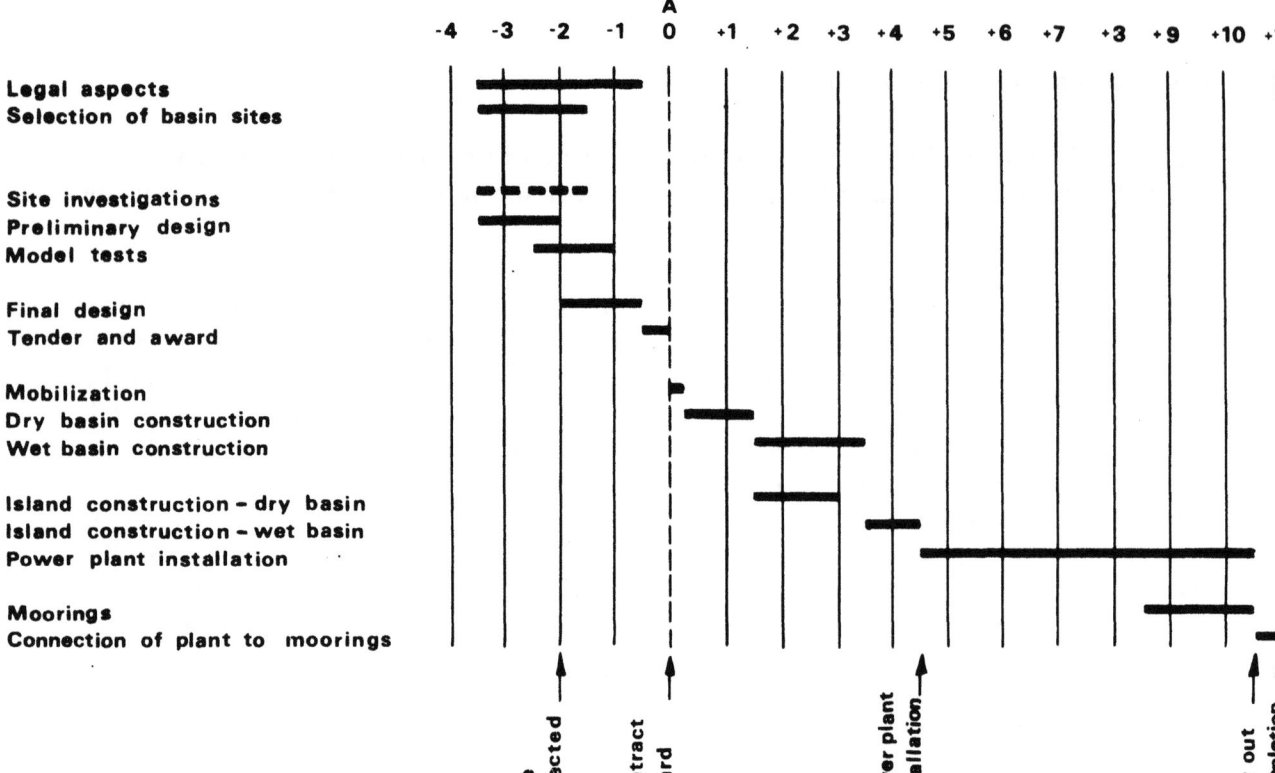

Floating island - construction programme for 12 ha area

Drawing 19

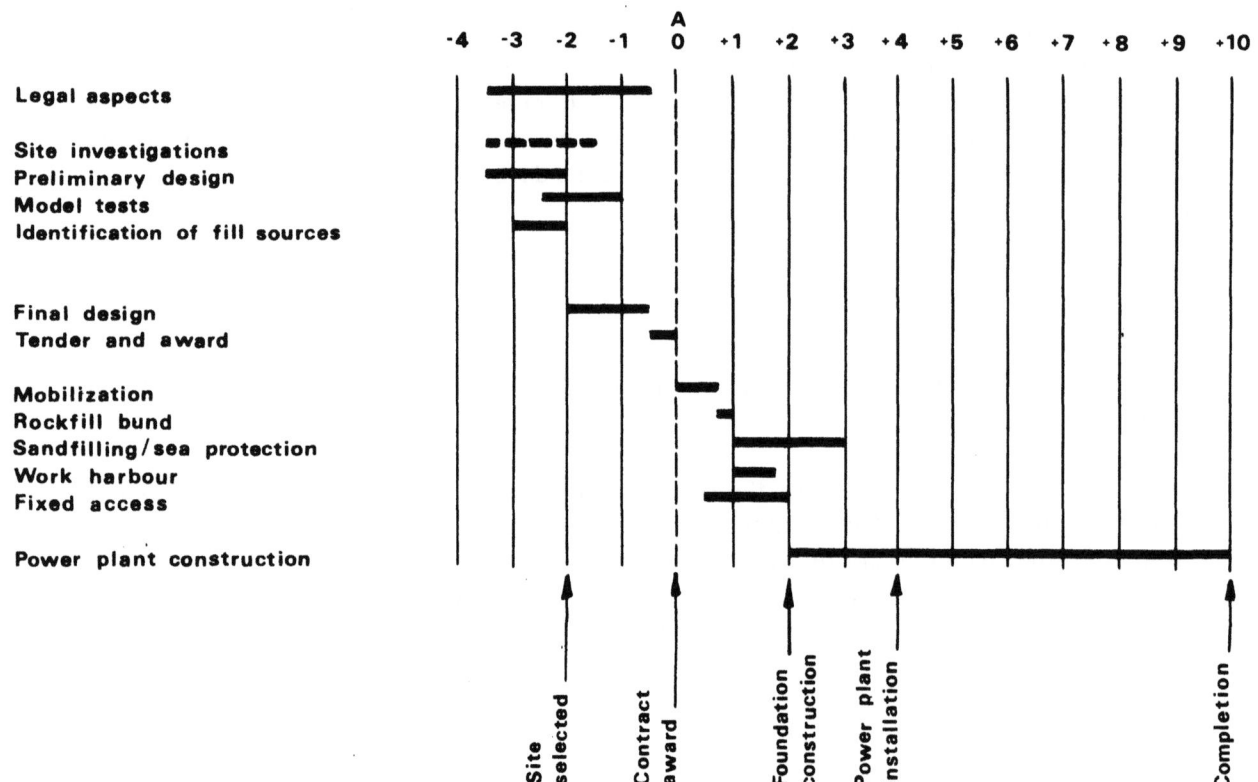

Fill island - construction programme for 12 ha area

Drawing 20

7.10   These problems are illustrated in Drawing 18 which gives a plot of maximum speed against frontal area of an island for various total tug powers. The practical limit to the number and size of tugs means that a maximum of not more than 75 MW or 100,000 HP total can be employed. With this total power the maximum towing speed for the largest islands will be low.

Positioning

7.11   The positioning of the island and the connection of the mooring ropes present further problems. In order to reduce the connection time the ends of the moorings must be connected in groups to floating vessels. When the island is towed out the vessels to which the groups of mooring cables are attached will be connected to the island.

Overall programme

7.12   A possible outline programme for the construction of a floating or semisubmersible island is shown in Drawing 19. Times are given in years from the date of the award of the contract for the island construction, marked as year A.

7.13   Site investigations at the possible basin sites and at the offshore mooring sites would start immediately the initial authorization to proceed was given — (A-3.5). Preliminary designs would also start at this time. These would be followed by model testing of alternative designs, selection of the preferred design and final design.

7.14   The construction of the dry basin would be started immediately after the award. When this was completed, (A+1.5), the island construction and construction of the caissons for the wet basin would start in the dry basin. Preliminary work on the wet basin could also start at this time. On completion of construction in the dry basin (A + 3) the caissons for the wet basin would be floated out first and towed to their final location. The partially completed island would be floated out some six months later when the wet basin was complete. Power plant installations would start a year afterwards (A + 4.5). Tow out would be 10½ years after award. Once at its operating site the island would be connected to the moorings which would have already been installed.

7.15   This programme could be shortened by a year if basin construction were let as a separate contract and started before the award of the contract for the island construction. In this case power plant installation could start 3½ years after award of the island contract (A + 3.5) and float out would be 9 years after award.

Single basin construction

7.16   In this method all construction work is carried out in a dry basin. Island and plant construction are completed before the structure is floated and towed out. Although this method avoids the need to build an expensive wet basin and permits plant installation to be carried out at a more accessible coastal site it has two serious disadvantages.

7.17   Because of the increased weight of a completed island deep water is needed for the float out from the dry basin. A minimum of about 25 m of water is required. The cost of constructing a dry basin to give this depth of water will be much greater than that to provide the 8 m draught required for two basin construction.

7.18   As all the construction takes place in a dry basin the island cannot be ballasted down to maintain the area of construction activity at sea level. As work proceeds it will need to be carried out at progressively higher levels and the nuclear plant will be built at at the highest level. In the case of the largest semisubmersibles this means building a plant at about 40 m above the general ground level. This will increase the cost and slow the rate of plant construction.

7.19   Single basin construction does not seem likely to produce a shorter construction period than two basin construction. The programme for two basin construction shows that building a wet basin is not a critical activity. Its elimination does not reduce the overall construction time.

7.20   The cost of the deeper dry basin needed for single basin construction and the extra cost of installing the plant on an elevated platform seem likely to offset any savings of this method. The deep channel required for tow out from the dry basin can only be found at a few places and dredging of a special channel will increase costs. Although no wet basin is required some deep water site for ballasting down a semisubmersible before tow-out to its operating position is desirable. If a semisubmersible is not ballasted before being towed out to the open sea it will be floating with the full length of its legs above sea level. In this state a semisubmersible is vulnerable to wave damage. However, this method has been adopted in suitable weather conditions for some North Sea structures.

7.21   Improved stability during tow out and a lower deck level during plant installation for a semisubmersible could be achieved by some form of jack-up structure. In this case the deck soffit would be close up to the raft during plant installation and float out. On arrival at the operating site the deck would be raised up from the raft on telescopic legs or legs accommodated in pockets in the deck. The fully extended structure would then be ballasted down to its operating freeboard. The feasibility of such an operation involving the raising of nearly 1 Mt from a floating base is doubtful. It is unlikely that even if practicable such a complex operation would produce any cost savings over simpler methods.

Modular construction

7.22   In this approach the main raft is divided into a number of smaller units for construction. The need for one large dry basin is eliminated and several smaller basins, possibly in different locations, would be satisfactory.

7.23   The reduced size of each section means that it may be possible to use existing construction yards or shipbuilding facilities to build the modules. This presents potential cost and time savings.

7.24   If modules are built in conventional ship yards rather than in a purpose built dry basin, float out by flooding the construction area may not be possible. Launching of raft modules from a slip way is an alternative. The largest sections of raft envisaged for modular construction are about 285 x 50 m. Such units will weigh between 100,000 t and 200,00 t with draughts of between 8 m and 15 m depending on how much of the raft is constructed before launching. Thus the sections being considered are comparable with the largest bulk carriers or tankers.

7.25   After launch or float-out the modules will be towed to a wet basin where they will be connected together. Satisfactory jointing of floating units is a particular problem of this approach. Completion of island construction and plant installation then proceeds as for the dry basin/wet basin method.

Double dry basin construction

7.26   This method, which is only applicable to semisubmersibles, will use two large dry basins simultaneously but will not require an elaborate wet basin. However, some deep water construction site is needed for assembly of the sections floated out from the dry basins.

7.27   The raft and legs will be constructed in one dry basin and the deck and plant in a second dry basin by the methods described in Section 7.6. Access for construction of the nuclear plant will be simpler than if it is built in a wet basin.

7.28   When the raft and legs are completed they will be towed to the deep water site. To float out the raft and legs a water depth of about 15 m is needed in the dry basin. At the deep water site the raft will be ballasted down until the legs are beneath the water surface. This effectively involves sinking the structure and if sea bed conditions are suitable the raft will be sunk to rest on the sea bed. If the water is too deep or bottom conditions are unsatisfactory the raft and legs will be ballasted down until they are suspended from special pontoons built for this purpose.

7.29   The nuclear plant will then be towed to the deep water site using the deck section for temporary buoyancy. The combined plant and deck require a draught of about 10 m. This is equal to the depth of deck needed to support the plant when finally installed above water level. To provide adequate buoyancy during float out the sides of the deck section will be temporarily raised by attachment of prefabricated units.

7.30   Once the plant and deck are positioned above the legs, the raft and legs will be raised. To achieve a controlled docking of the two parts of the island the raft will either be maintained at neutral buoyancy and be jacked up or be deballasted and allowed to rise against cables fixed to the sea bed. When the legs and deck are docked, ballast can be pumped out to raise the whole structure to the required level for towing to its operating site.

7.31   This method appears attractive becasue it removes the need for extensive construction facilities at a wet basin. By proceeding with island and plant construction in parallel it might be possible to save up to 1½ years on the time to completion compared with the programme for wet and dry basin construction. However, the site needed for assembling the raft and deck sections needs to be deep enough to allow the legs to be lowered at least 10 m below sea level. A semi-submersible designed for a wave (H_D) of 6 m height would need a minimum depth of water of 55 m at the deepwater site to be constructed in this way.

Summary

7.32   The table below shows the main characteristics of the alternative construction methods considered.

Comparison of construction methods for a semi-submersible designed for H_D = 6 m

Method	A Wet & dry basin	B Single basin	C Modular	D Double dry basin
Area of dry basin (mxm)	300 x 300	300 x 300	6 No 50 x 300	2 No 300 x 300
Minimum depth of dry basin(s) (m)	8	25	8	15 and 12
Minimum depth of wet basin/assembly site (m)	45	Not needed	45	55
Plant installation	In wet basin	At high level in dry basin	In wet basin	At low level in dry basin
Time for completion (from award of contract) (yrs)	10½	10½	10½	9

7.33   A major problem with all the proposed methods is the need for deep water. The maps in Appendix E show the depths of water available along the coastline of the EEC.

7.34   Method B does not require the extremely deep water sites used for wet basin construction or assembly in the other methods. However the dry basin for method B is the deepest of any scheme and would be very costly to build. The nuclear plant has to be installed at a high level which is unattractive.

7.35   Method D needs the deepest water although extensive construction facilities do not need to be built at this site. The use of two dry basins reduces the construction period but increases the cost of mobilisation. The leg and deck sections, which have to be connected offshore, are larger than any structures which have been manoeuvred in this way to date.

7.36   Methods A and C have similar water depth requirements. Method C is only likely to be preferable to Method A if existing shipbuilding or North Sea Platform facilities can be used. The feasibility of building and launching large concrete structures from shipyards is uncertain.

Summary

7.37   Method A, wet and dry basin, which uses techniques already proved in the building of large concrete production platforms for the North Sea, is the most acceptable construction method at the present time. Construction facilities for this method are costed in Section 9. The tow-out of the vessel will only be feasible if the construction basins are in areas without large tidal currents. The connection of the mooring ropes will present further problems on a scale not yet attempted.

Gravity Platform

General

7.38   The construction methods available for building gravity platforms are basically similar to those for semisubmersible islands. The wet and dry basin method is again preferred for the reasons summarised above. However, gravity platforms for deep water operation are even larger than semisubmersible structures. Their increased weight means either that deeper dry basins are needed or that less work can be completed in a dry basin before float out. A greater water depth is also required at the wet basin. The actual depth depends on the water depth at the final location. Towing difficulties are increased due to the large size of these structures. The method of positioning is described below.

Positioning

7.39   The site for the base of the fixed platform will have to be levelled prior to the tow-out. This operation will require excavation to tight tolerances in deep water. The operation of ballasting the platform down onto the foundation will require a weather window without appreciable wind or waves so that the structure can be accurately positioned. Once the platform has been ballasted down it may be necessary to grout the area beneath the base to ensure intimate contact.

Caisson Island

7.40   The construction methods for caisson islands are similar to those for floating islands. The caissons will be constructed in dry basins and the plant installed in wet basins. The caissons will be towed out individually and positioned in a similar manner to a gravity platform.

7.41   The programme for the construction of a caisson island will be similar to that for a floating island shown on Drawing 19.

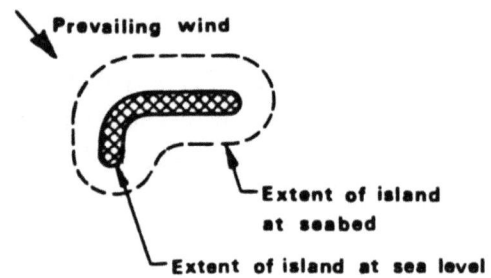

Stage 1

Placing of rockfill bund for wave protection

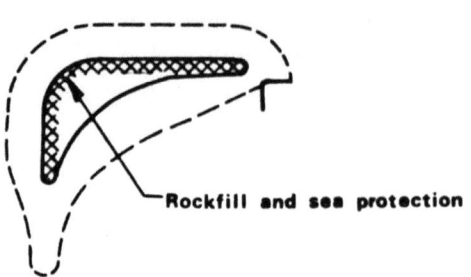

Stage 2

Placing of rockfill bunds to retain sandfill
Placing of first sandfill
Start of harbour construction

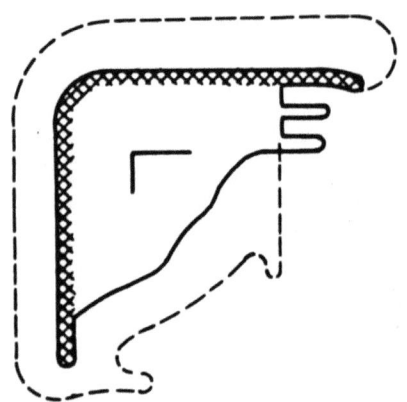

Stage 3

Harbour in use
Power plant construction starting
Placing of rockfill, sandfill and sea protection continuing

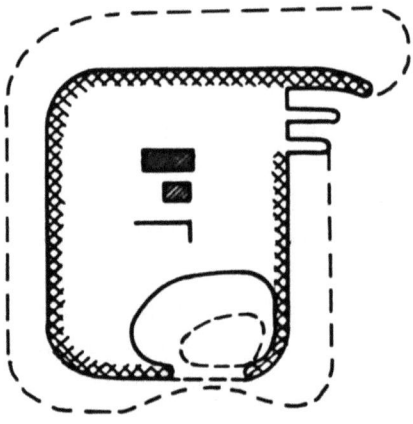

Stage 4

Power plant construction continuing
Rockfill bund being closed prior to completing sandfill

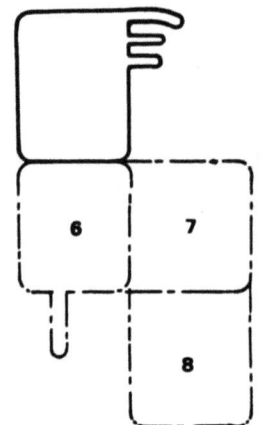

Stage 5

Completion of first two 1250 MW power plants to give 2500 MW capacity

Development to 5000, 7500 and 1000 MW shown as 6, 7 and 8. The location of these later stages of development will depend on site conditions

Construction sequence for fill island   Drawing 21

Fill island

General

7.42   The construction of the fill island must be arranged so that the construction of the power station can be started as soon as possible. In this way the overall financing costs will be minimized. Sufficient filled area and permanent access facilities must be established at an early date.

7.43   To achieve a rapid build-up of construction activity, a work harbour at the island must be given priority. As well as securing the access to the growing island, a work harbour reduces the need for shore facilities.

Construction sequence

7.44   The construction sequence we envisage is shown on Drawing 21. The initial stage of construction will involve constructing a rockfill bund on the seabed on the more exposed side of the site. This bund, which will be armoured as it is built, will be raised above sea level. This gives protection for the hydraulic filling which can start in its lee. The length of the rockfill bund will depend on the local topography and sea conditions. In very exposed locations a row of caissons might be preferable to a rockfill bund. In very calm locations or enclosed water the rockfill bund may be unnecessary.

7.45   The area of hydraulic filling can be extended behind the rockfill bund, the extensions being contained by a series of rockfill bunds forming open lagoons at successively higher levels.

7.46   The work harbour can be built either using the sandfill as a temporary working platform or prior to filling using marine plant. In the former case diaphragm walling and tied back or cellular sheet piling can be used. In the latter case caissons, concrete block work or sheet piling can be used. The choice will be determined by the particular site needs. If the work harbour is to be used as the permanent harbour with an 80 year design life then the concrete options, although perhaps slower to construct, are preferable.

7.47   During these early stages of construction extensive use will be made of marine plant including bottom dump barges for placing the underwater sections of the bunds and riprap, and jack-up platforms for placing the dolosse units. Underwater trimming to profile will be carried out by barge mounted hydraulic back-hoes fitted with rakes. Sophisticated position fixing apparatus will be used.

7.48   The method of winning the sandfill will depend on the locations of the borrow areas. If these are within a few kilometres of the island, a cutter-suction dredge pumping directly to the site could be used. However, it is more likely that sea conditions and distances will favour the use of trailer dredges either dumping directly or feeding a reclamation dredge which can pump the sandfill up above sea level.

7.49   As the filling progresses and a fixed access, if chosen, is completed, work can start on the foundations for the power station. A wellpoint dewatering system or diaphragm walls will be required.

7.50   The final closure of the external bund may involve placing areas of bed protection to prevent scour if a high tidal range exists.

Overall programme

7.51   An outline programme is shown on Drawing 20 for a 12 ha fill island in Category I or II with fixed access being constructed to the shore. Times are shown in years + or - the date of the award of contract (A).

7.52   Immediately the possible sites have been identified (three and a half years before the award of the contract, A-3.5) a preliminary site investigation can be carried out to determine the most attractive ones. These investigations will include boring the sea bed and collection of the data on currents, littoral drift etc which will be required for the model testing.

7.53   The results of these investigations will be used to determine the favoured site and to enable model testing to be started. Physical models have a long lead-in time.

7.54   At A-2 the preliminary design and economic analyses of the various options will be complete and sufficient data available from the model tests to confirm the actual site and the shape of the island. At this stage the detailed site investigation and final design can be started. Tenders will be called 1½ years later (A-0.5).

7.55   The length of the mobilization period will be determined by the time taken to develop the rockfill sources and to set up the shore facilities. These will be kept to a minimum. The first construction work on site will be the placing of the rockfill bund although the fixed access can be started earlier from the shore. After two years (A+2) the fixed access will be complete and sufficient sand fill placed for the construction of the first power plant to start. Sand filling and sea protection would be finished a year later (A + 3).

7.56   The construction time for a 30 ha island in Categories I or II could be similar to that for a 12 ha island if say two 4,000 m^3 trailer dredges were used instead of one. The slope protection for a 30 ha island could be placed in two years by two crews allowing 30% downtime. An overall construction of three years is possible. For a fill island in deeper water, say 20 m, the overall programme would be similar except an additional year should be allowed for sand filling before the start of power station construction.

Caisson retained island

7.57   The construction method for the caissons will be to construct them in a dry basin and then float them out and sink them onto prepared foundations using the methods already described.

7.58   The filling cannot start until an initial line of caissons has been placed to provide shelter for the filling operation to proceed without wave attack or scour. Thus the construction of the caissons is on the critical path. The mobilization and construction of the dry basin could be achieved 2½ years after the award of the island contract (A+2.5). Thus filling would start at A+2.5 instead of at A+1 for a fill island. Thereafter the programme would be similar to that for the fill island but delayed by 1½ years.

Protected floating island

7.59   The floating island will be constructed as described in Section 7.2 and the breakwater on the same principles as the sea protection for a fill island.

7.60   The critical path for the construction will be through the construction of the floating island. The overall programme will be similar to that for a floating island.

Caisson/fill and caisson/piled islands

7.61   The caissons will be constructed as described in Section 7.2. The fill of the caisson/fill island will be constructed as a fill island. The piled perimeter of the caisson/piled island will be constructed initially using marine plant and subsequently from travelling rigs supported on the piles already driven.

7.62   The critical path for both types of island will be through the caisson construction. The overall programme will be similar to that for a floating island.

Power station development

General

7.63   This section covers briefly the ways in which a 2,500 MW station could be extended to 5,000 MW and 10,000 MW.

Floating and semisubmersible islands

7.64   A development of 2,500 MW would require two individual islands constructed concurrently. To avoid congestion of their moorings, the two islands would have to be moored apart. Auxiliary islands would likewise have to be moored separately.

7.65   The extension of the development to 5,000 and 10,000 MW would require the addition of further islands all moored separately. Although some auxiliary islands might be shared, the extension will require the duplication of the main islands with little potential for saving.

Gravity platform

7.66   A development of 2,500 MW would require two individual gravity platforms with smaller platforms for ancillary equipment. The platforms could be positioned sufficiently close to each other to allow bridges to be built between them.

7.67   The extension of the development to 5,000 and 10,000 MW would require duplication of all the main platforms - some smaller auxiliary platforms could be shared. The potential for saving is therefore limited.

Caisson island

7.68   A development of 2,500 MW would require two sets of caissons each supporting a 1,250 MW power station and ancillary equipment. The caissons could be positioned adjacent to each other.

7.69   The extension of the development would require duplication of all the main platforms with only some smaller auxiliary platforms being shared.

Fill islands and caisson retained islands

7.70   One possible way that a fill island could be extended is shown on drawing 21. A caisson retained island could be extended in a similar way. There is some potential for saving if decisions can be made at an early date on whether the extensions are to be constructed.

Composite islands

7.71   The caisson/fill and caisson/piled islands can be extended in a similar fashion to a fill island development. The protected floating island can only be extended by complete duplication unless a very early decision is made to allow for extension by enlarging the area enclosed by the breakwater.

8. PLANT CONSTRUCTION COSTS

General

8.1   The total cost of a nuclear plant on an island can be split into three parts:

>   cost of power plant
>   cost of island
>   cost of ancillaries.

8.2   This study is primarily concerned with the cost of the island and some of the ancillaries. The selection of unit costs for estimating the island cost is discussed in Section 9. Estimates of the costs of islands and ancillaries using these rates are given in Section 10. However, the type of island, its method of construction and its final location will have a significant effect on the cost of the nuclear power plant itself. As the cost of the power plant* may be as much as ten times that of the island, a small percentage change in the power plant cost will have a large effect on the overall economics of the choice of island. In order to highlight the possible effects on the power plant construction, we discuss below some of the factors involved for the different types of island.

Floating island

8.3   The plant can be installed at the deep water basin adjacent to land in which the island is completed. The number of sites available for deep water basins of the 40 m draught required are few. They may well be remote from manufacturing and residential areas which will increase the cost of plant construction compared with a conventional land based power station.

8.4   The island can be ballasted to bring its deck down to the level of the adjacent wet basin quay. However, the cost of providing the quay will be high and all round access may be uneconomic. Restricted access and confined working areas will increase the plant costs. Constructing the island in sections and joining them together after plant installation might reduce access restrictions.

Semisubmersible island

8.5   The same factors apply to the semisubmersible island as to the floating island.

8.6   The draught of the deep water basin required for the semisubmersible island is more, say 45 m, and the problems are more severe. The possibilities for constructing the plant in sections are more limited.

---

* *A typical land based power station costs £1000M per 1000 MW.*

Fixed island

8.7 The installation of the power plant on a fixed island is a major problem.

8.8 If the power plant is to be installed in the deep water basin, the island will have to be fully ballasted to maintain stability. The water depth required will be of the same order as the water depth at its final location. We are considering depths of 50 to 100 metres. Sites for deep water basins of 50 m or more depth will be difficult to find. Possible locations are limited to a few areas in Europe. If they can be developed they are likely to be remote with associated increased costs of transport and labour.

8.9 If the power plant is installed after the island is fixed in its final position, the access problems are likely to be severe. We do not consider that this method of installation is a feasible alternative.

Fill island

8.10 A fill island gives the opportunity of constructing the power plant either conventionally in situ or by prefabricated elements remote from the island site. In the first case ease of access will be the critical factor in determining the cost of the plant. For inshore locations a fixed access in the form of a bridge or a tunnel could be provided. Where the island is remote from the shore the advantages of prefabricating the heavier elements, such as the reactors, could be significant. This method minimises the the amount of work to be carried out in an exposed location and enables closer control of fabrication standards to be exercised. An extension of this approach to include prefabrication of all the plant is the caisson/fill island. However, construction of dry basins for the caissons delays overall completion of the plant and island. The provision of extra platform area as construction areas on the island is likely to produce an overall benefit for most sites. The exact size of the most cost effective platform will be site dependent. We have taken a 30 ha island for illustrating the fill island type (Section 6.70).

8.11 Foundation and dewatering problems will be site specific but will not necessarily be any more severe than for coastal stations.

Caisson and composite islands

8.12 The power plant is installed on a caisson in the same way as floating islands. The comments in Sections 8.3 - 8.5 apply to the composite islands as well.

Summary

8.13 The effect of these factors on the cost of the nuclear plant and the design modification that may be possible to produce the most cost effective solution are beyond the Terms of Reference of this study. Study of these factors could form part of any further work undertaken as an extension of this study.

8.14 In general, for the optimum sites, the cost of the power plants will be greatest for the fixed islands and least for fill islands; semisubmersible and floating islands being between these two.

9. UNIT COSTS

General

9.1　The estimation of the costs of the islands is based on summing the products of the estimated quantities of principle materials and the assumed unit costs. Certain other costs described below have been added as appropriate. To this total an allowance of 25% has been added for contingencies and 7% for engineering.

9.2　The unit costs have been derived from our experience of work of a similar nature. They reflect the difficulties and hazards of maritime construction and in particular the problems of island construction.

9.3　The unit costs are dependent to varying degrees on the location of the island and the shore facilities. However, as this study is not site specific, it has not been possible to determine unique unit costs. The unit costs given represent average figures for the individual materials. We have assumed that the location of the islands would be chosen so that, as far as possible, the overall cost of materials is minimised.

9.4　The base date for the unit costs is January 1980.

Concrete in islands

9.5　The costs of islands constructed from concrete have been considered under four headings:-

>an "all-in" rate for the concrete construction
>the cost of building construction basins
>the cost of tow-out
>the cost of foundations and moorings

9.6　The different types of island have been costed on the basis of an "all-in" rate of £450/m^3 of concrete. This is equal to £600/m^3 after the addition of contingencies and engineering. The rate of £450/m^3 is based on works of a similar, but smaller, nature in the North Sea and elsewhere. It has been adjusted to allow for the particular difficulties of the size of the structures. The "all-in" rate includes for all materials, labour and overheads necessary for constructing the island. The cost of the basic island structure is the volume of concrete in the island multiplied by the "all-in" rate.

9.7　To this figure must be added the cost of the construction basins. This can be a significant proportion of the total cost. Sites for the basins will require careful selection. The wet basin requires deep water, 25 m minimum, close to land with good communications with the surrounding area. For the construction of a 2,500 MW station the wet and dry basins are assumed to be duplicated so that two 6 ha, 1,250 MW units can be constructed in parallel. £20M has been added to the cost of a 6 ha floating island or caisson island to allow for the construction basins. £40M has been added to the cost of a 6 ha semisubmersible or gravity platform for the construction basins.

9.8   The cost of tow-out is partially dependent on the distance which the island has to be moved from the construction basin to its operating location. Thus this cost is a site specific element. Based on a towing distance of 500 km and the curves of required tug power shown on Drawing 18 we have added 1% to island costs to allow for tow-out.

9.9   On arrival at their operating site fixed platforms will be ballasted down to the seabed. The gravity structures for which outline designs have been prepared do not require the heavy piled footings necessary for legged jackets. We have assumed that sites will be chosen where the seabed is level and the in-situ material is able to support the structure. When the structure has been ballasted down to the seabed grouting will be required under the foundations. We have added 3% to the cost of fixed platforms to allow for this work.

9.10   Floating structures need extensive moorings to restrict the accelerations of the power plant. As discussed in Section 6 the feasibility of such large moorings is in doubt. However, to assess the order of magnitude of mooring costs a system of tensioned steel cables fixed to concrete anchor blocks has been considered. In estimating the costs of moorings a rate of £1,000/t has been used for steel cable.

Sandfill

9.11   Under the most favourable conditions sandfill can be won and placed for as little as £0.60/m^3. Such a unit cost could only be achieved if the fill were won adjacent to the island and sea conditions allowed the uninterrupted use of a large cutter-suction dredger discharging by pipeline directly into the island. In practice such ideal conditions are unlikely to pertain and a higher unit cost is more realistic. If sea conditions and haul distances require the use of trailer dredgers then, for haul distances of 10 to 15 kilometres, a unit cost of £0.85/m^3 is realistic. However, the material to be placed, above the level at which the trailer dredgers can dump, will have to be rehandled by suction dredgers. This will increase the cost by some £0.40/m^3 for the rehandled fill. We have therefore taken the average unit cost for sandfill as £1.25/m^3. We feel that this is a reasonable figure for an island in categories I or II and for large islands in category III for which large sea-going trailers can be used economically.

9.12   The unit cost will increase pro-rata to the days lost for more exposed locations and at a rate of approximately £0.30/m^3 for each 10 km additional haul.

Natural gravel

9.13   Natural gravel may be used either for filter material or for an artificial beach.

9.14   Sea-bed deposits of gravel are generally small and fairly scattered. Larger deposits are found in deep water (below 35 m) but these could be difficult to extract economically. Considerable haul distances may be involved in procuring gravel. In addition the dredging operation must be carefully controlled in order to try to obtain an acceptable grading without the need for reclassification of the material. The basic cost of gravel delivered to the site from a source 50-100 km distant will be around £5.50/m^3. If the material is used as filter then the cost of placing will increase the total cost to £10.00/m^3. For an artificial beach the placing cost will be lower and a unit cost of £6.25/m^3 has been taken.

9.15   The unit costs represent a compromise. The cost of placing the material above water level is reduced because land plant can be used to spread the gravel and thicknesses can be determined with greater confidence. However costs are increased as the gravel cannot be dumped directly into place. Below water level the material can be dumped into place but spreading is difficult and greater thicknesses are required to ensure continuity of the layer.

Rockfill

9.16   The major elements in the total cost of quarry-run rockfill are transporting and handling. If a quarry could be opened with direct access to a jetty then a unit cost in the range of £3.00-£4.00/m^3 would be taken. In practice environmental considerations will probably preclude the opening of a new quarry and existing quarries may have to be used. Thus, allowing for land transport and double handling a unit cost of £7.00/m^3 has been taken.

Filter

9.17   Crushed rock from 150 mm down will be required as a filter material. The material will have to be quarried and crushed on land and then shipped to the island. Using current (UK) quarry prices and a sea haul distance of 200 km gives a cost delivered to the site of £9.00/m^3. Allowing the same costs for placing as used for the gravel makes a total unit cost for the crushed rock of £14.50/m^3.

9.18   Rock between 500 mm and 150 mm will be required as a coarse filter material. The rock could be produced by passing quarry-run rockfill over a 150 mm sieve. The small material would pass on to the crushers for aggregate production and the override used as the coarse filter. Alternatively it could be selected from the waste material produced during production of the larger sizes of rock armour. A unit cost of £14.50/m^3 has been taken.

Rock armour

9.19 The rock armour has been subdivided into "derrick stone" for use beneath precast concrete armour units and "rip rap" for use on its own. The differences in grading are unlikely to make significant differences to the unit costs of materials having the same D_{50} size. The D_{50} size of particulate material is the length of the side of a square opening through which 50% by weight of the material will pass. Unit costs have been built up for ranges of D_{50} sizes.

9.20 The production of rock armour in large sizes and large quantities can cause severe environmental problems both around the quarry and along the land haul route. The handling of the material is expensive and rehandling must be kept to a minimum. It may be cheaper to quarry in a remote location where environmental problems are less severe and rehandling can be minimized. These savings may more than offset the cost of a sea haul of even 2000 km. This overseas option creates an upper bound to the range of unit costs.

9.21 The smaller sizes of rock armour (up to 3t) can be handled by large front-end loaders but large sizes may require cranes. The unit costs taken reflect the higher costs of handling the large sizes. They also assume that there is a market either elsewhere or at the offshore island for the smaller sizes of stone produced in the quarrying operation.

Concrete in armour units

9.22 Concrete armour units will require to be mass produced. The unit costs of producing armour units will decrease as the size of block increases. However, this is counteracted by the increased difficulty and cost of placing large armour blocks. We have therefore taken a constant rate, irrespective of size, of £55.00/m^3 of concrete, which includes for casting, handling and placing armour units.

Sheet piled quays

9.23 The harbours have been costed on the basis of the length of quay wall required with an allowance for the protective breakwater. The rates taken for sheet piled quays are £2,500/m and £10,000/m for quays in 5 m and 10 m depth of water respectively.

Summary

9.24 The unit costs taken are set out in Table 5.

9.25 To calculate island costs, estimated quantities have been multiplied by the unit costs shown in Table 5.

9.26 Further amounts have been added for construction basins, tow-out, moorings and foundations as appropriate.

9.27 Finally an allowance of 25% has been added to these costs for contingencies plus 7% for engineering to give a total island cost.

TABLE 5 UNITS COSTS OF CONSTRUCTION MATERIALS
(EXCLUDING CONTINGENCIES AND ENGINEERING)

Material	Unit	Cost (£)
Concrete in islands	m^3	450.00
Steel in moorings	t	1000.00
Sandfill	m^3	1.25
Natural gravel for beach	m^3	6.25
Natural gravel filter	m^3	10.00
Rockfill	m^3	7.00
Crushed rock filter	m^3	14.50
Rock armour, D_{50} size 0.15-1t	m^3	18.00
" " " " 1 -3t	m^3	22.00
" " " " 3 -10t	m^3	25.00
" " " " 10t	m^3	28.00
Concrete in armour units	m^3	55.00
Sheet piled quays 5 m water depth	m	2,000
10 m water depth	m	10,000

TABLE 6 Costs of floating and fixed islands

Floating Island - 2500 MW (2 x 6 ha)

H_D (m)		3	6	12
Cost of structure	(£M)	330	500	880
Cost of moorings in 50 m	(£M)	60	160	500
Cost of moorings in 100 m	(£M)	80	220	710
Total cost in 50 m	(£M)	390	660	1380
Total cost in 100 m	(£M)	410	720	1590

Semisubmersible island - 2500 MW (2 x 6 ha)

H_D (m)		3	6	12
Cost of structure	(£M)	840	920	1000
Cost of Moorings in 100 m	(£M)	70	150	290
Total cost in 100 m	(£M)	910	1070	1290

Gravity platform - 2500 MW (2 x 6 ha)

H_D (m)		3	6	12
Total cost in 50 m	(£M)	1000	1060	1140
Total cost in 100 m	(£M)	1360	1460	1530

Caisson island - 2500 MW (2 x 6 ha)

H_D (m)		3	6	12
Total cost	(£M)	340	500	840
Draught of reactor caisson	(m)	25	30	40

10. COST ESTIMATES

General

10.1 Using the unit rates from Section 9 and allowances for construction basins, tow-out, moorings and foundations as appropriate we have costed our outline designs for the following types of Island:-

        A   floating island (concrete)
        B   semisubmersible island (concrete)
        D   gravity platform
        G   caisson island
        K   fill island
        L   caisson retained island
        N   protected floating island
        P   caisson/fill island
        Q   caisson/piled island

10.2 An allowance of 25% has been added to the total for contingencies and a further 7% for engineering.

10.3 The figures derived represent average costs. The unusual nature of some of the structures under consideration and the dependence of many costs on site conditions introduce a degree of uncertainty into the costings. This should be borne in mind when interpreting the costs quoted in this Section.

10.4 The costs are simple total costs and not net present values (NPV's). For the purpose of general comparison the simple total costs are sufficient to show the relative rankings of the types of islands (the NPV's are of the order of 85% of the simple total costs for most types of island). For similar reasons the maintenance costs are not included. Study of NPV's and maintenance costs would form part of any further studies in which a few selected types of islands would be examined in greater detail.

10.5 The effect on island costs of reductions in the weight and area of nuclear power plants is discussed in sections 10.27 - 10.34. The costs of providing additional area for auxiliary structures, access to the island and transmission are estimated separately in section 10.35 onwards.

Costs of outline designs

Floating island

10.6 A 2500 MW development consisting of two 6 ha islands each supporting a 1250 MW power plant has been costed for wave heights (H_D) of 3, 6 and 12 m. The costs are shown in Table 6. For comparison with fill islands the estimate assumes that the two islands are constructed at the same time. Duplication of the wet and dry construction basins has been allowed for. The costs of these construction facilities are given in Section 9. Sequential construction would reduce the cost of construction facilities.

10.7   The feasibility of producing a satisfactory mooring system for a floating island is in doubt. Approximate costs have been calculated assuming that the moorings consist of steel cable fixed to concrete anchor blocks on the seabed. Costs for moorings in 50 m and 100 m depths of water and the combined costs of the islands and their moorings are shown in Table 6.

10.8   For developments to plant sizes greater than 2500 MW costs increase almost pro rata. Some saving can be made because the original construction basins can be re-used for later plants. This reduces the cost of each further 2500 MW installation by about £40 M. Other comparatively small savings should arise as construction techniques are refined with experience.

Semisubmersible

10.9   The cost of a 2500 MW development consisting of two 6 ha islands is shown in Table 6 for a range of wave heights. The estimates assume duplication of the wet and dry construction basins.

10.10   The same type of mooring system has been assumed for semisubmersibles as for floating islands. The lower costs of moorings for semisubmersibles are a result of the reduced wave forces on these structures. The costs of moorings and the combined cost of semisubmersibles and moorings are shown in Table 6.

10.11   The saving on construction basin costs for development beyond 2500 MW amounts to about £80 M for each subsequent 2500 MW installation.

Gravity platform

10.12   The cost of a 2500 MW development comprising two 6 ha platforms is shown in Table 6 for water depths of 50 and 100 m. Wave heights of 3, 6 and 12 m have been considered. Duplicated wet and dry basins have been included in our estimates at a cost of £80 M.

10.13   Cost reductions similar to those for semisubmersibles will arise if construction basins are reused to produce additional platforms.

Caisson Island

10.14   The cost of a caisson island for a 2500 MW development is shown in Table 6. As discussed in Section 6.64 this type of island is best suited to a water depth equal to the draught of the deepest caisson. These depths are shown on Table 6. In deeper water fill can be placed below the caissons. Beyond the practical limits of fill placing the caissons themselves must be extended. This would cost about £12 M for every metre extension. In water shallower than the draught of the caissons, dredged foundations and approach channels would have to be formed. These costs would vary with the site conditions.

Fill Island

10.15　Costs have been calculated for fill islands with three alternative types of wave protection:

>　　Dolosse concrete armour units on 1 on 2 slope
>　　Rip rap on 1 on 4 slope
>　　Sandfill beach on 1 on 50 slope

10.16　The capital costs of 12 ha and 30 ha islands protected with dolosse and rip rap are compared on Drawing 22. For design waves in excess of 3 m the large size of rock required for stability makes riprap impractical. Even for 3 m waves a rip rap protected island is more expensive than a dolosse protected island. The difference becomes more marked as water depth increases.

10.17　We estimate that the present value of maintenance costs over the design life of a dolosse or rip rap protected island would amount to about 10% of the construction cost.

10.18　The capital costs of 12 ha and 30 ha islands protected with sand beaches are shown on Drawing 23. These costs are considerably higher than those for the alternative types of sea protection.

10.19　Maintenance costs are also likely to be higher although their value would depend on site specific conditions such as tides and currents. For a typical beach we estimate that the present value of replenishment costs over the design life would be at least 30% of the construction cost.

10.20　Wave protection formed from dolosse or other interlocking concrete armour units appears to be the most economic solution over the range of wave heights and water depths considered.

10.21　Fill islands can readily be enlarged to accommodate the addition of further power plants. If development is envisaged from the outset the cost per unit area of forming an island can be reduced as the size is increased. This is because the protected perimeter of the island, which is a major part of the cost, need not be increased in proportion to the area formed. The way in which costs vary for developments of 2500 MW to 10000 MW are shown in Table 9.

Caisson retained island

10.22　The costs of a 12 ha caisson retained islands for a plant installation of 2500 MW are shown in Table 7. This type of island is more expensive than a fill island with concrete armour block protection.

Protected floating islands

10.23　A 2500 MW development consisting of two 6 ha islands within a protecting breakwater has been costed. A design wave of 3 m within the breakwater has been used.

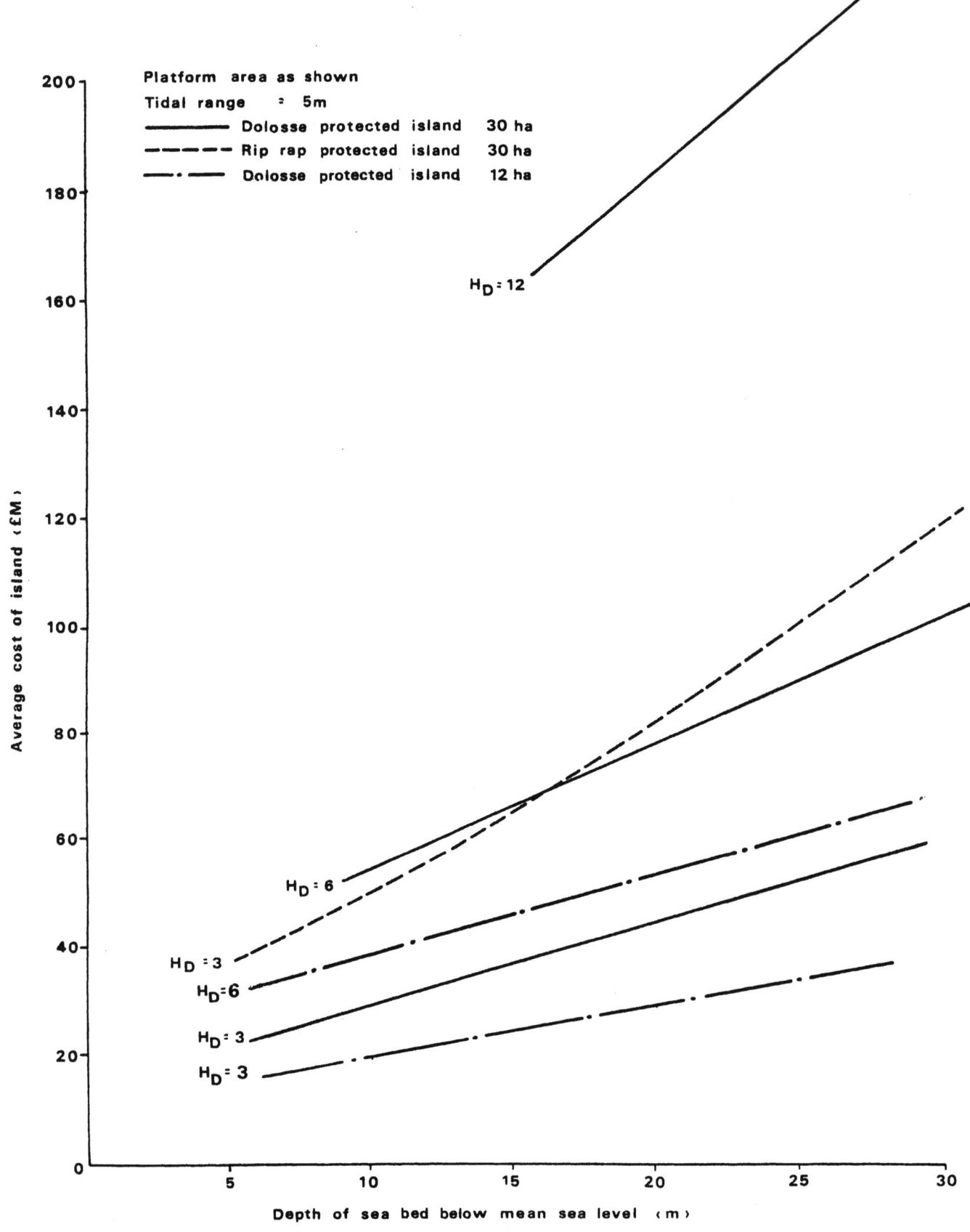

Comparison of cost of islands protected with dolosse and rip rap

Drawing 22

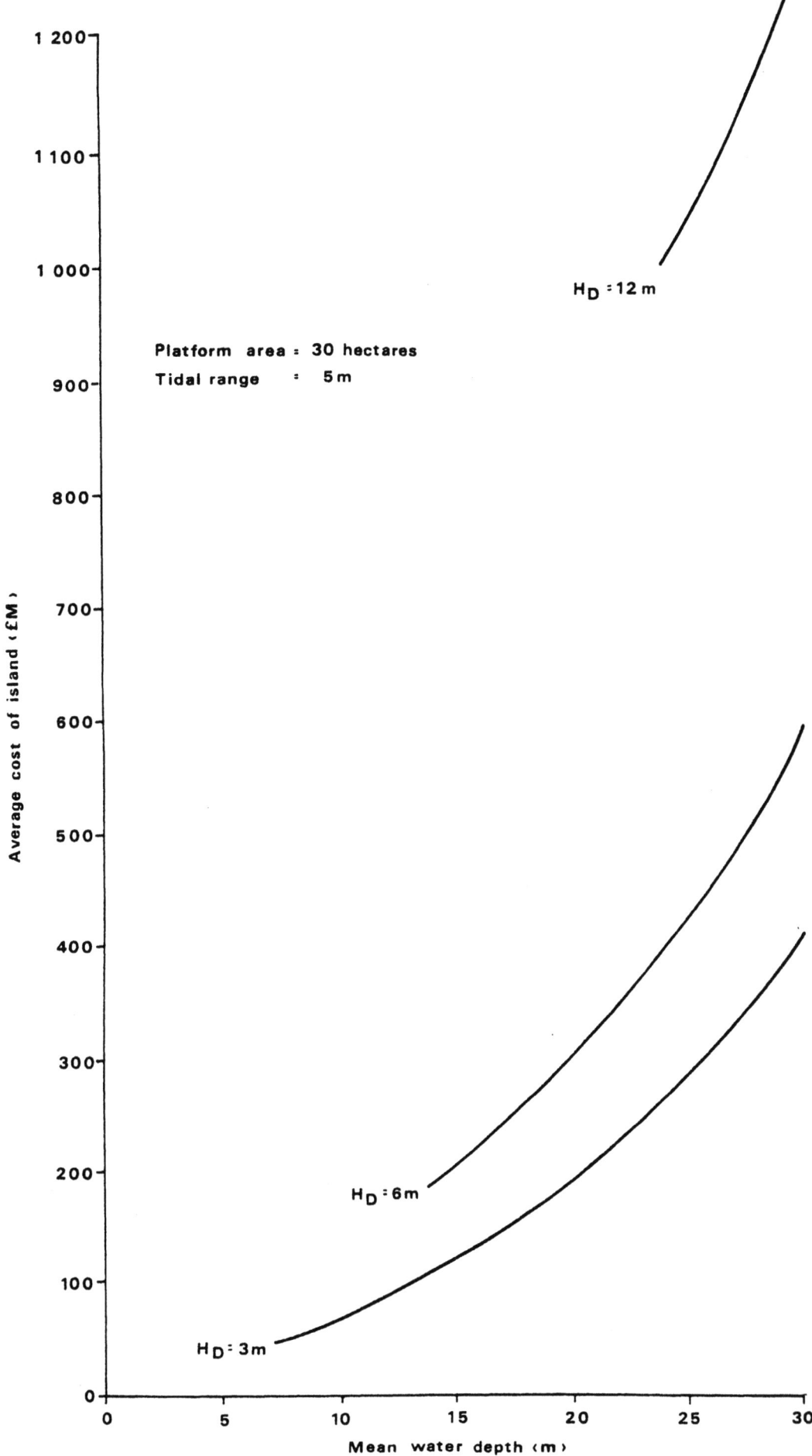

Capital cost of island protected with sandfill beach at 1 on 50 Drawing 23

TABLE 7 Costs of fill and caisson retained islands 2500 MW development

12 ha fill island

	Costs (£M)		
H_D (m)	3	6	12
10 m mean water depth	20	40	—
20 m mean water depth	30	50	140
30 m mean water depth	40	70	180

30 ha fill island

	Costs (£M)		
H_D (m)	3	6	12
10 m mean water depth	30	60	—
20 m mean water depth	50	80	190
30 m mean water depth	60	110	240

12 ha caisson retained island

	Costs (£M)		
H_D (m)	3	6	12
10 m mean water depth	50	60	—
20 m mean water depth	100	130	—
30 m mean water depth	170	220	290

30 ha caisson retained island

	Costs (£M)		
H_D (m)	3	6	12
10 m mean water depth	70	90	—
20 mean water depth	150	200	—
30 m mean water depth	240	310	460

10.24 The costs of the breakwaters enclosing the two floating plants have been calculated for wave heights of 3, 6 and 12 m outside the breakwater. The combined cost of the islands, moorings and breakwaters are shown on Table 8.

10.25 Unless provision is made at the time the first breakwater is built, expansion to beyond 2500 MW will involve duplication of the breakwaters and islands. As for floating islands a saving of about £40 M can be made on each further 2500 MW unit by reusing the wet and dry construction basins.

Caisson/fill and caisson/piled islands

10.26 The costs for a 2500 MW development of each of these types of composite island are shown on Table 8. The caisson/fill island has been costed for an area of 30 ha which is comparable with a fill island. The caisson/pile island costs are for a 12 ha development. We have assumed that auxiliary structures would be used with caisson/pile islands, as with floating and fixed islands, to give additional space.

10.27 The costs of caisson fill islands increase with water depth in the same way as for fill islands. The costs of caisson/pile islands increase more rapidly because of the large piles required in deep water. In water depths of less than about 25 m, dredging will be needed for both types of island for float in of the caissons.

Sensitivity of island costs to changes in nuclear plant size

10.28 The power plant weights on which our outline designs and costings have been based are those given in the draft information note (Appendix B). These weights are summarized in Table 2 in Section 4. The redesign of the nuclear plant to reduce weight and size is beyond the Terms of Referencee of this study. However such redesign has been carried out in other studies of floating nuclear power stations. To enable an assessment to be made of the savings that may be derived from such redesign, we have investigated the sensitivity of the island costs to reductions in nuclear plant weights and sizes.

10.29 The effect of reducing the weight of the nuclear plant is most marked for floating islands. The table below compares the costs of floating islands designed to support plants of 430,000t, 200,000t and 100,000t. The plant area in all three cases is assumed to be the same and the island in each case measures 245 x 245 m. Some cost reduction results from reducing the plant weight but the savings are a small proportion of the total cost, particularly when the design wave height (H_D) is large.

Variation of floating island cost with plant weight

Power plant weight (t)	H_D (m)	Cost of structure (£M) 3	6	12
430,000		350	530	920
200,000		230	420	790
100,000		200	380	750

(Island area 6 ha for all plant weights)

TABLE 8 Costs of composite islands 2500 MW development

12 ha Protected floating island

H_D (m)	Costs (£M)		
	3	6	12
20 m mean water depth	430	460	—
30 m mean water depth	480	510	590

12 ha caisson/fill island

H_D (m)	Costs (£M)		
	3	6	12
10 m mean water depth	320	—	—
30 m mean water depth	330	350	460

30 ha caisson/fill island

H_D (m)	Costs (£M)		
	3	6	12
10 m mean water depth	320	—	—
30 m mean water depth	350	390	520

12 ha caisson/piled island

H_D (m)	Costs (£M)		
	3	6	12
10 m mean water depth	320	440	—
30 m mean water depth	350	390	850

10.30   Further cost reductions arise if the plant area is reduced as well as its weight. The table below compares costs of floating islands of different areas supporting power plants of 100,000t weight. Costs decrease markedly with a reduction in island area even for large design waves.

Variation of floating island cost with plant area

Island area (ha)	H_D (m)	Cost of structure (£M)		
		3	6	12
6		200	380	750
4		160	290	550
2		130	200	310

(Plant weight 100,000t for all areas)

10.31   Reductions in plant weights and areas may be produced by refinements and adaptations of nuclear plant technology allowing the same 1250 MW power output to be produced from more compact units. Alternatively the costs shown above could be interpreted as the cost of islands to support small nuclear plants of conventional design but lower output.

10.32   Stability of small nuclear plants has not been considered as these are outside our terms of reference. However, smaller and lighter islands are likely to present dynamic stability problems under wave action.

10.33   Decreases in the costs of semisubmersibles and fixed islands with reductions in plant size would also occur. However, savings would not be so large as for floating islands. Fill islands would not be affected by changes in plant weights although the costs of foundations for the nuclear plants would be reduced for lighter installations.

10.34   Even for the lightest and smallest plants considered the cost of a floating island remains higher than a fill island for water depths up to 40 m. The costs of other island types are unlikely to change so much due to reductions in plant weight that they will become cheaper than fill islands.

10.35   Other considerations than cost may favour small floating islands. The mobility of this type of unit allows it to be used for temporary applications. The smaller draught required for lighter islands widens the areas available for construction sites.

Auxiliary structures

General

10.36   In addition to the main plant building, which for each 1250 MW power plant will occupy 6 ha, further space must be provided for auxiliary structures. These include such items as switchgear, workshops, administrative buildings, helicopter landing pads and staff accommodation.

TABLE 9 Costs of auxiliary structures and staged development

Semisubmersible (load 50 kN/m^2)

H_D	Cost/hectare (£M)		
	3	6	12
50 m mean water depth	60	75	110

Caisson

H_D	Cost/hectare (£M)		
	3	6	12
20 m mean water depth	25	35	50
30 m mean water depth	35	45	60

Piled deck (load 50 kN/m^2)

H_D	Cost/hectare (£M)	
	3	6
10 m mean water depth	10	20

Fill island - stage development in 20 m depth

H_D (m)	Costs (£M)	
	3	6
2500 MW	40	70
5000 MW	70	130
7500 MW	100	190
10000 MW	140	260

10.37   The ways in which this space might be provided for the different island types is discussed below.

Floating and fixed islands

10.38   Enlargement of the deck areas of floating structures or gravity platforms will increase the cost of these structures approximately in proportion to the area increase. The auxiliary structures are much lighter than the main plant buildings and are likely to be able to tolerate more movement than the main plant. Shorter design lives may also be approximate for some auxiliary structures. It is, therefore, likely to be more economic to support the auxiliary buildings on separate lightweight structures rather than adding to the size of the already large main structures.

10.39   The costs of semisubmersible platforms to support an average deck load of 50 kN/m^2 are given in Table 9.

10.40   It may not be necessary to install all the auxiliary items on separate structures. Considerable space exists within the cellular concrete decks of the floating islands and fixed platforms. This can be utilised to house many of the auxiliary items.

10.41   The costs of extending islands using caisson or piled additions is also shown in Table 9.

Fill islands

10.42   The outline designs for fill islands, caisson retained and caisson/fill islands have been based on 30 ha islands and costs are presented on tables 7 and 8. These tables also give the costs of 12 ha islands of these types to enable direct comparison with the other designs. As noted in Section 6.70, we consider that the increased costs of nuclear power plant construction that would arise if a 12 ha island were used, would be greater than the additional cost of constructing a 30 ha island which would include construction areas.

10.43   The protected floating island or caisson could be extended by providing a fill platform beside the island supporting the main plant. The costs of fill islands that could serve this purpose are shown on drawing 24 for a range of areas and site conditions. The costs shown assume that the full perimeter of the island has to be protected. They would be reduced if a fill platform is built alongside an existing structure such as the breakwater of the protected island.

Access

Fixed Access

10.44   The estimated costs of various fixed accesses are compared in Drawing 25 for varying water depth. The range of costs can be summarized as follows:

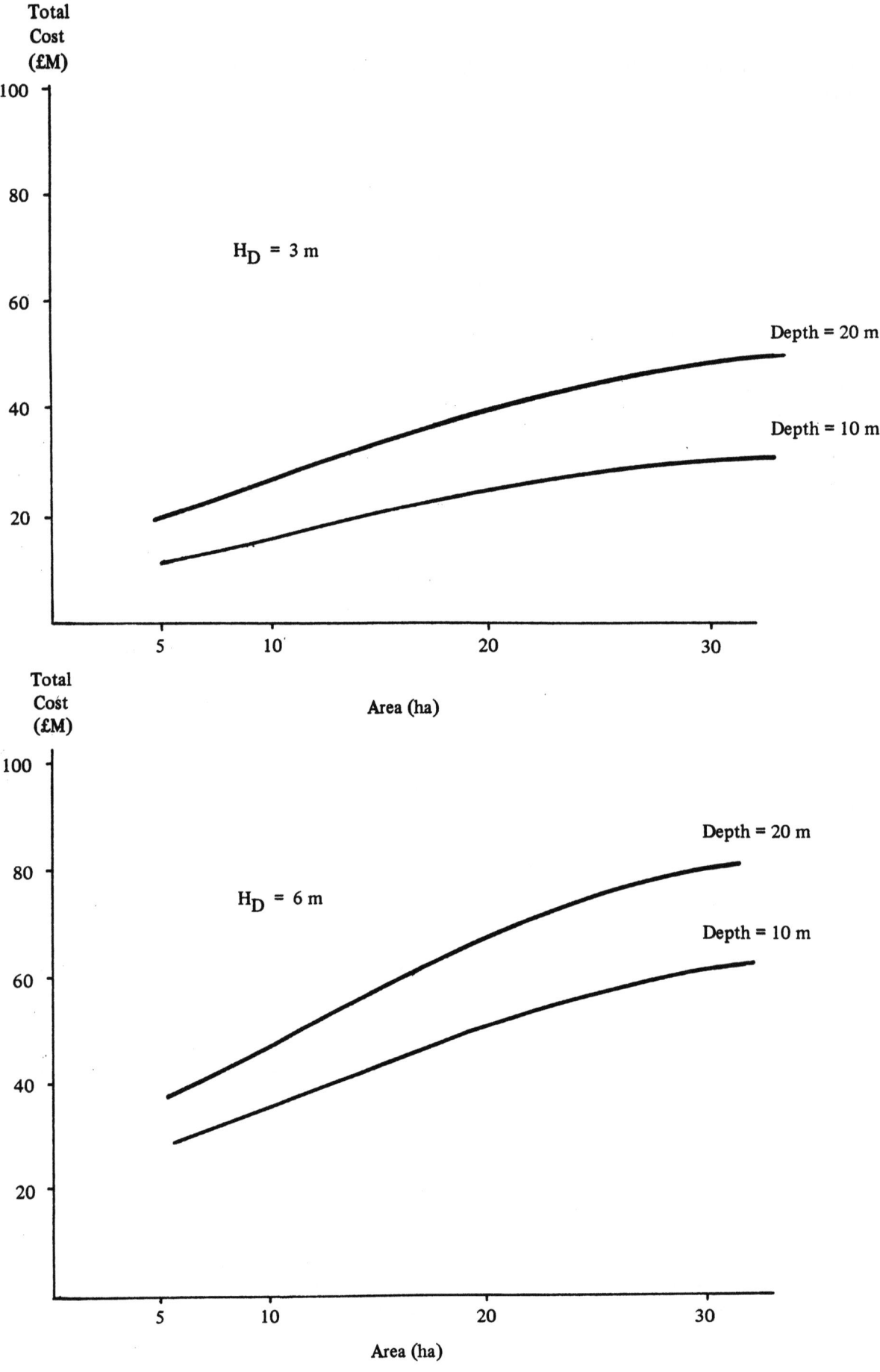

Variation of the cost of fill islands with area

Drawing 24

Type	Cost per km	
	Water depth 5 m	Water depth 30 m
Embankment	£ 15 M	£40 M
Bridges (single)	£ 6 M	£12 M
Bored tunnels (two)	£ 20 – £60 M for all depths	
Submerged tube (two)	£ 20 – £40 M for all depths	

10.45 The embankment costs are sensitive to the design wave height (H_D). The costs given relate to a design wave height of 5 m and a tidal range of 5 m. The cost will vary by approximately $\pm$ 15% for every 1 m change in H_D. The bridge costs are sensitive to foundation conditions. A dense sand foundation has been assumed for the piled piers. The range of bored tunnel costs indicates the effect of the geology.

10.46 It can be seen that the bridge access is the cheapest solution. An embankment will only be competitive in the shallowest depth considered. If the vulnerability of the bridge is considered unacceptable, it may be cheaper to duplicate the bridge rather than provide another type of access.

Sea access

10.47 If a fixed access is not provided, the cost of transporting the personnel and materials that would have used the fixed access, must be taken into account. The quantification of these costs is site specific and beyond the scope of this study. As an indication of the order of costs involved, the approx present value (PV) of providing a hovercraft or helicopter service to an island 25 km offshore are shown below. These costs are based on the numbers of operating staff listed in Section 4. The cost of transporting construction personnel would be similar to the figures shown.

	Hovercraft service PV cost (£M)	Helicopter service PV cost (£M)
2,500 MW development	30	70
5,000 MW development	55	120
10,000 MW development	100	210

Transmission links

10.48 The cost of submarine cabling has been taken as £3.5M per 1000 MW per km. This assumes the seabed material is soft. For the tranmission arrangements outlined in Section 4.83 the costs per km for the transmission links are:-

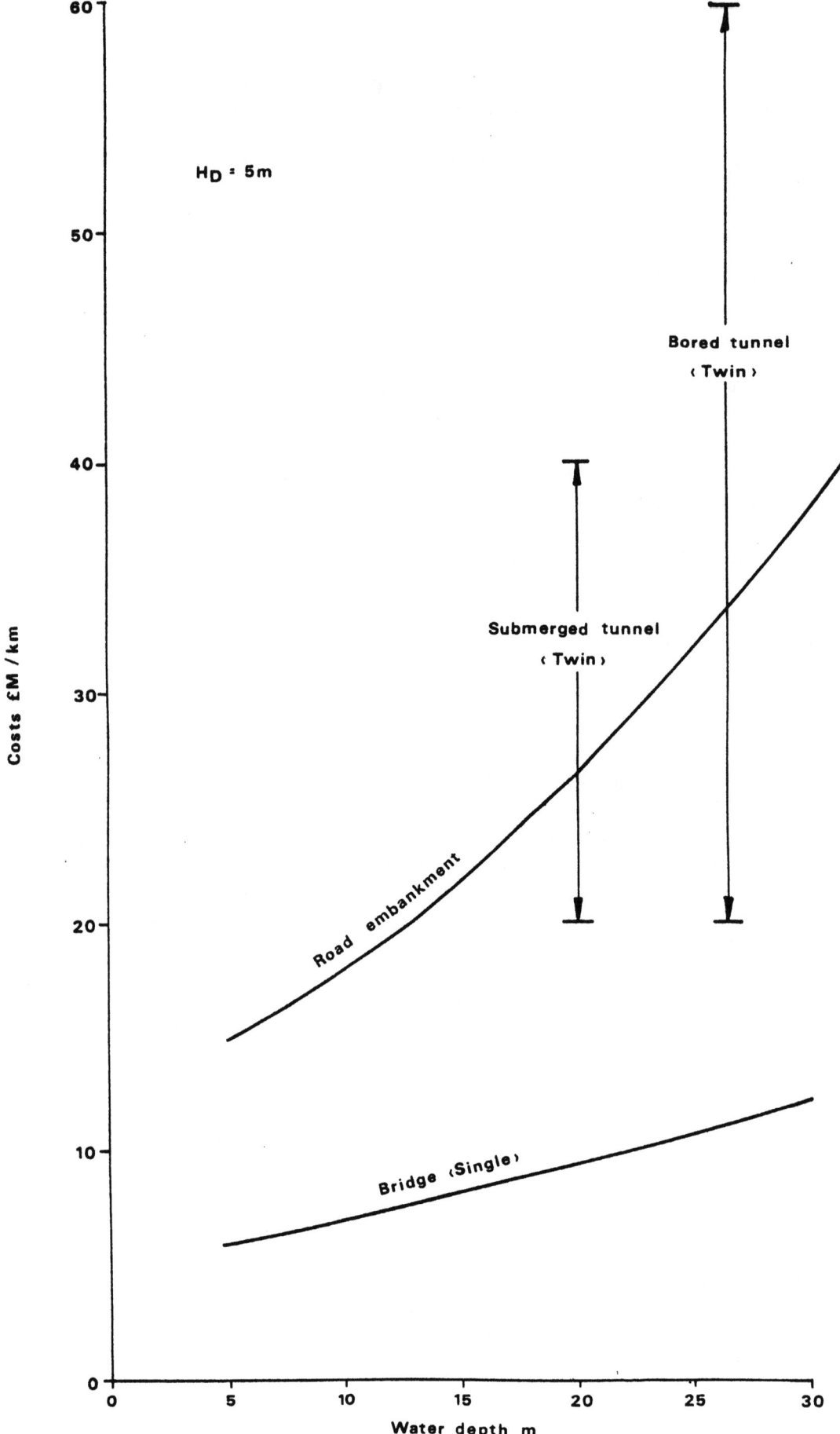

Costs of fixed access   Drawing 25

	No. of routes	Cost per Km (£M)
2,500 MW development	2	17.5
5,000 MW development	4	35.0
10,000 MW development	6	52.5

Summary

10.49   The total cost of a nuclear power station on an island can be split into three parts:

>   cost of power station
>   cost of island
>   cost of ancillaries

10.50   The cost implications of building a nuclear power plant on an island are discussed in Section 8.

10.51   The cost of ancillaries includes such items as auxiliary buildings, access and transmission. Auxiliary buildings will only form a relatively small part of the total cost. The costs of access and transmission are strongly site dependent.

10.52   A number of different types of island have been costed for a 2500 MW development over a range of water depths and wave heights. These costs do not include the cost of the power plant or ancillaries. The results of our calculations are summarised on Drawings 26, 27 and 28. These show the variation of cost with water depth for design wave heights (H_D) of 3, 6 and 12 m. The following points may be noted from these drawings:-

   (a) Technical considerations limit the range of conditions over which the different island types are feasible. For example semisubmersible islands will not float in water depths of less than 40 m while this is about the upper limit of the current experience of fill placing.

   (b) In water depths up to 40 m fill islands have the lowest costs for all the wave heights considered.

   (c) In deep water, beyond the current limits of fill placing, the cost of constructing offshore islands for nuclear power plants increases sharply.

   (d) Except for the most exposed conditions floating islands offer the cheapest solution for deep water. However the feasibility of mooring such structures to adequately restrict movement is seriously doubted.

   (e) In the most exposed conditions semisubmersibles offer the best solution. If floating islands are not feasible due to mooring problems semisubmersibles may also provide the cheapest structures for less exposed sites in water depths of more than 50 m. However, any structure in this depth of water is expensive.

(f) Gravity platforms and the various composite islands considered are generally more expensive than fill islands or unprotected floating structures.

(g) The costs refer to 12 ha, 2500 MW developments. The costs of extending the islands for auxiliary structures are discussed in Section 10.36–10.43 and shown in Table 9.

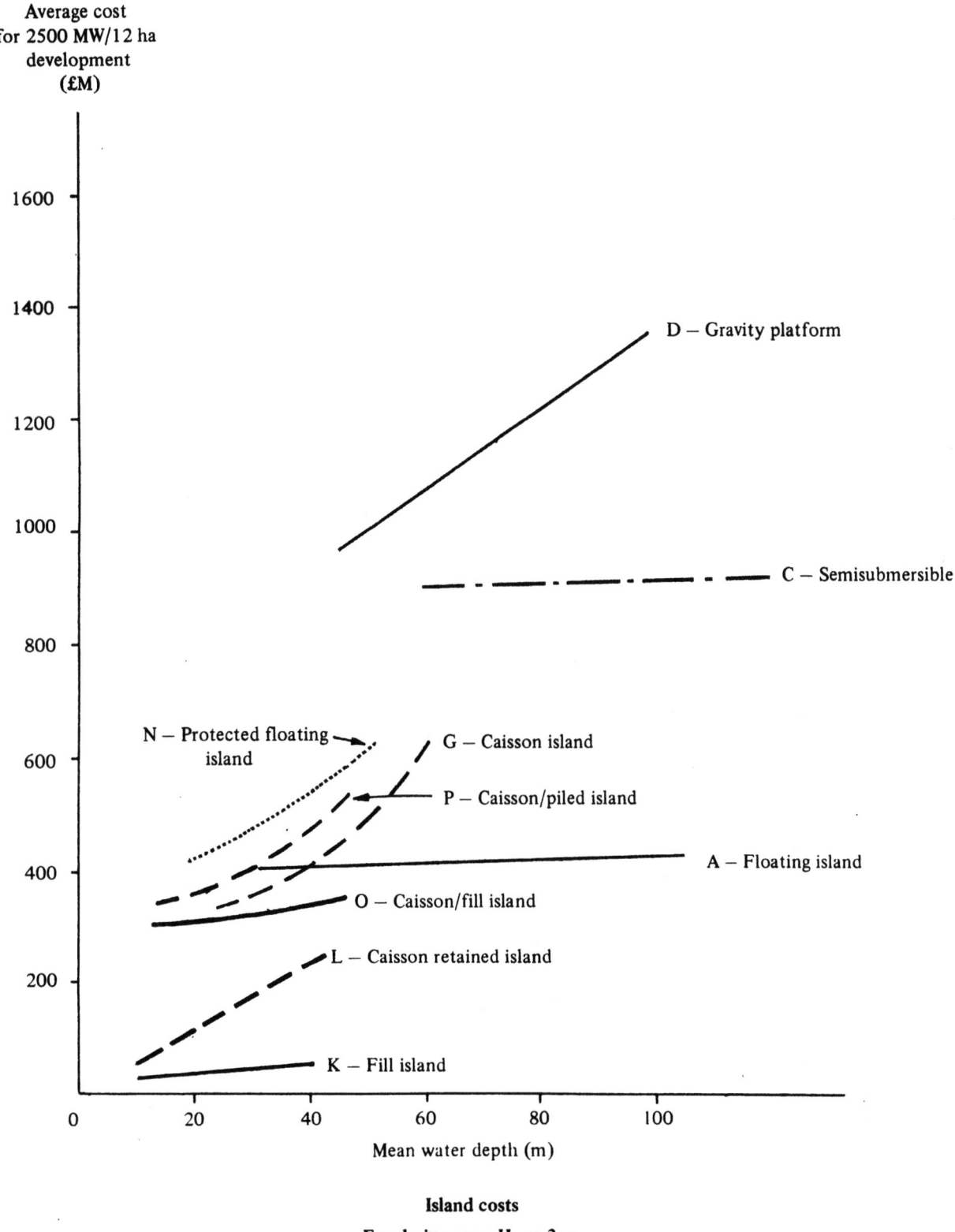

Island costs
For design wave H_D = 3 m

Drawing 26

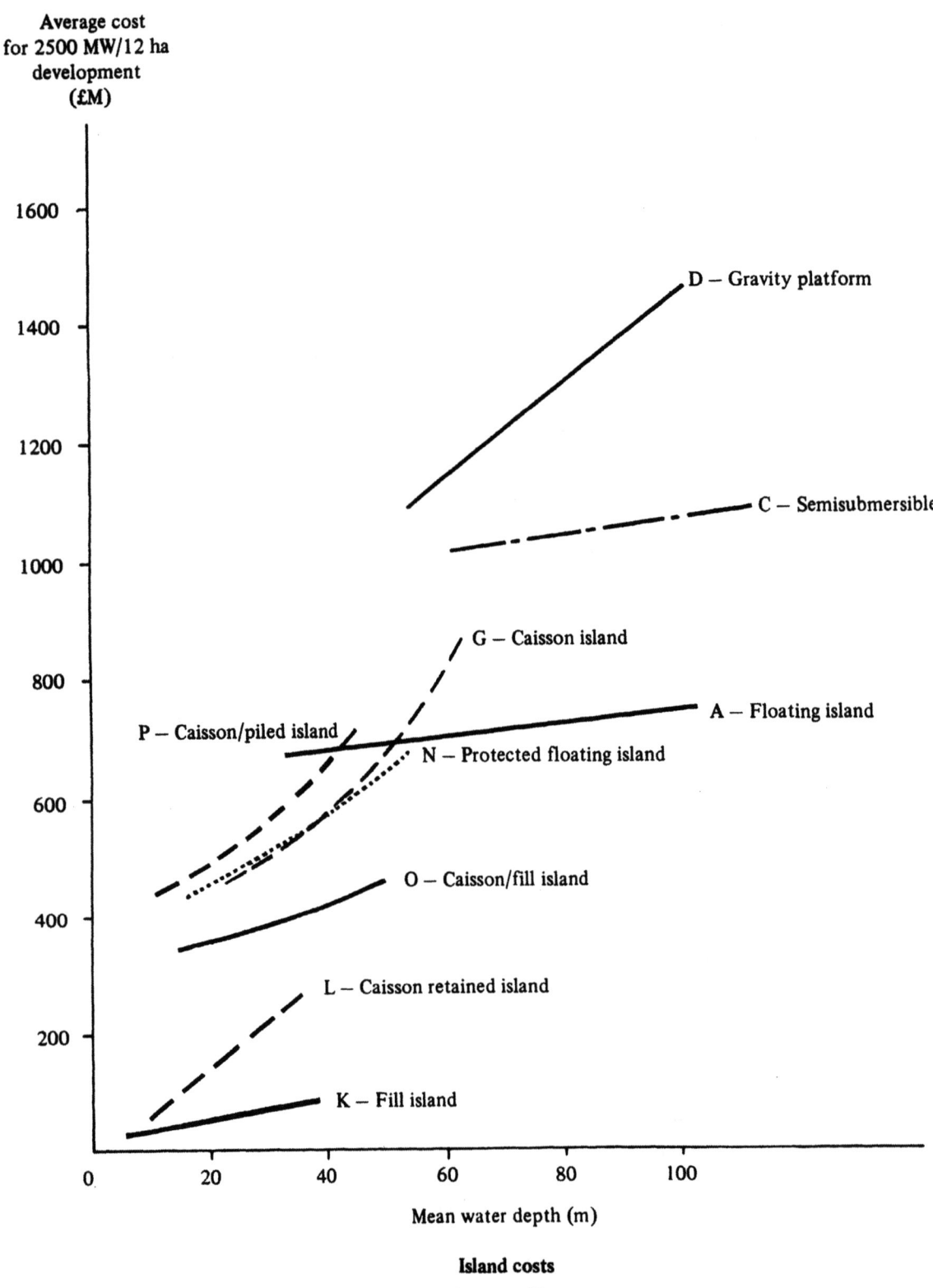

Island costs
For design wave H_D = 6 m

Drawing 27

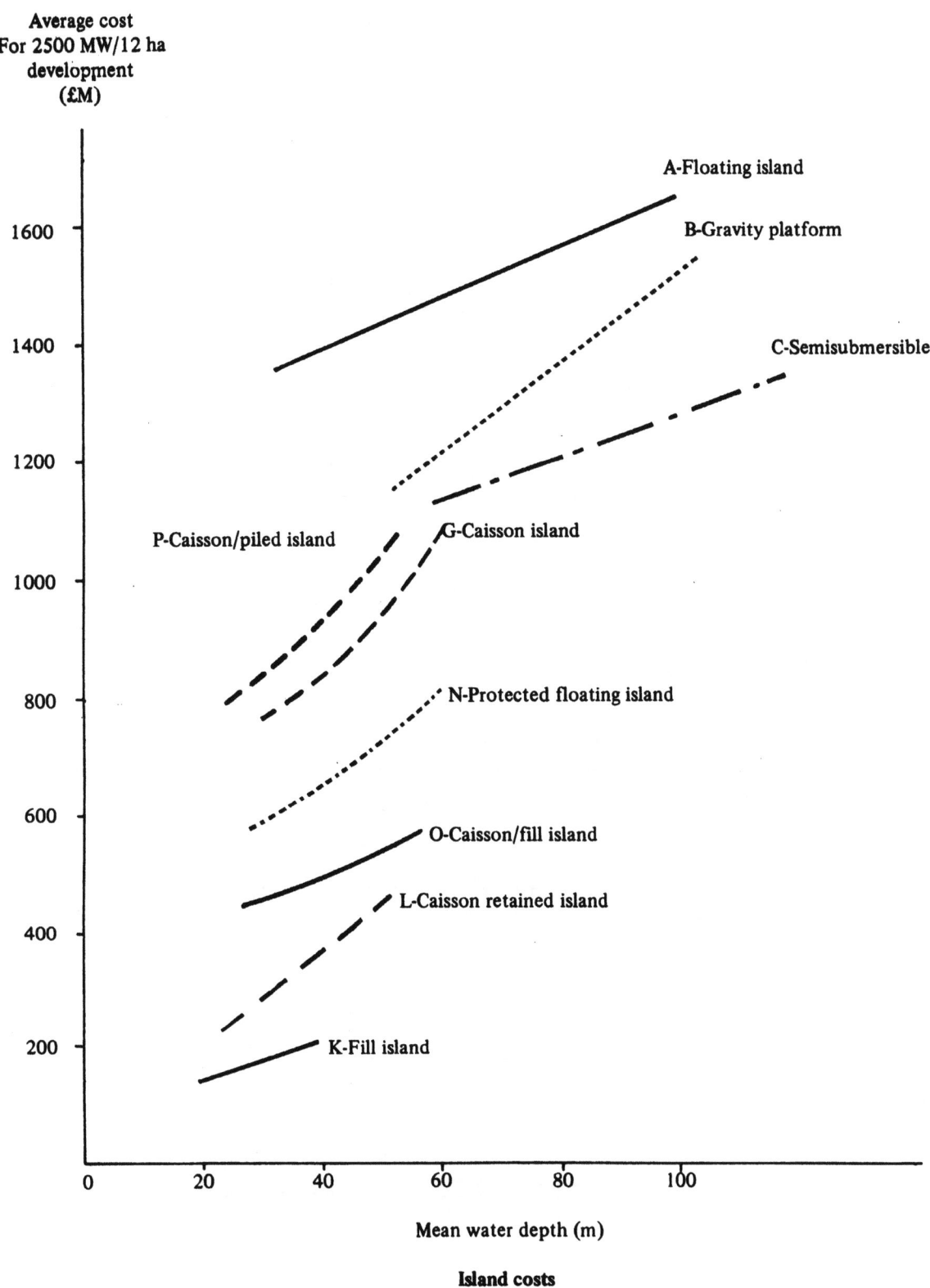

Island costs
For design wave $H_D = 12$ m

Drawing 28

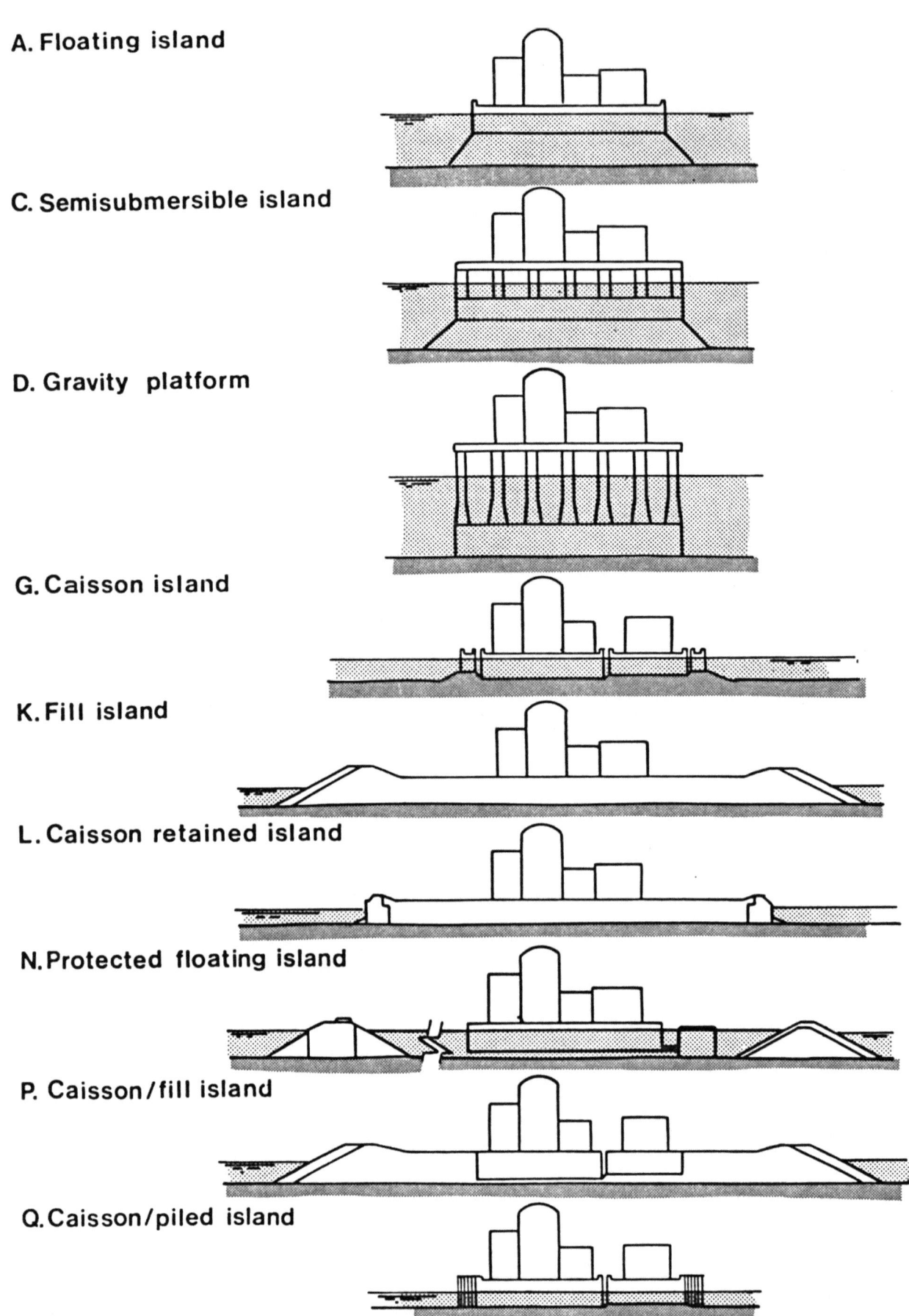

Preferred island types Drawing 29

11. CONCLUSIONS

Design Criteria

11.1 The design criteria to be adopted for an island supporting a nuclear power station must be very stringent. Agreed standards for installations of this type need to be established. Many design parameters are site specific and can only be established after a thorough programme of data collection, testing and analysis.

Types of Island

11.2 Many types of artificial island have been proposed most of them for use in oil exploration and production. Not all these types satisfy, or can be made to satisfy, the requirements of maximum safety, long design life and proven technology that are a prerequisite for an island supporting a nuclear power station. Further research and development may allow some of these findings to be amended.

11.3 The types of island that could be used and their main problems and difficulties are:

Type of island	Main problem or difficulties
A Floating island	— not feasible to moor in exposed locations
	— tow out
	— access
C Semisubmersible	— feasibility of mooring in exposed locations is in doubt
	— draught makes construction difficult
	— tow out
	— access
D Gravity platform	— installation of power plant
	— draught makes construction very difficult
	— tow out
	— access

TABLE 10 ISLAND SUITABILITY

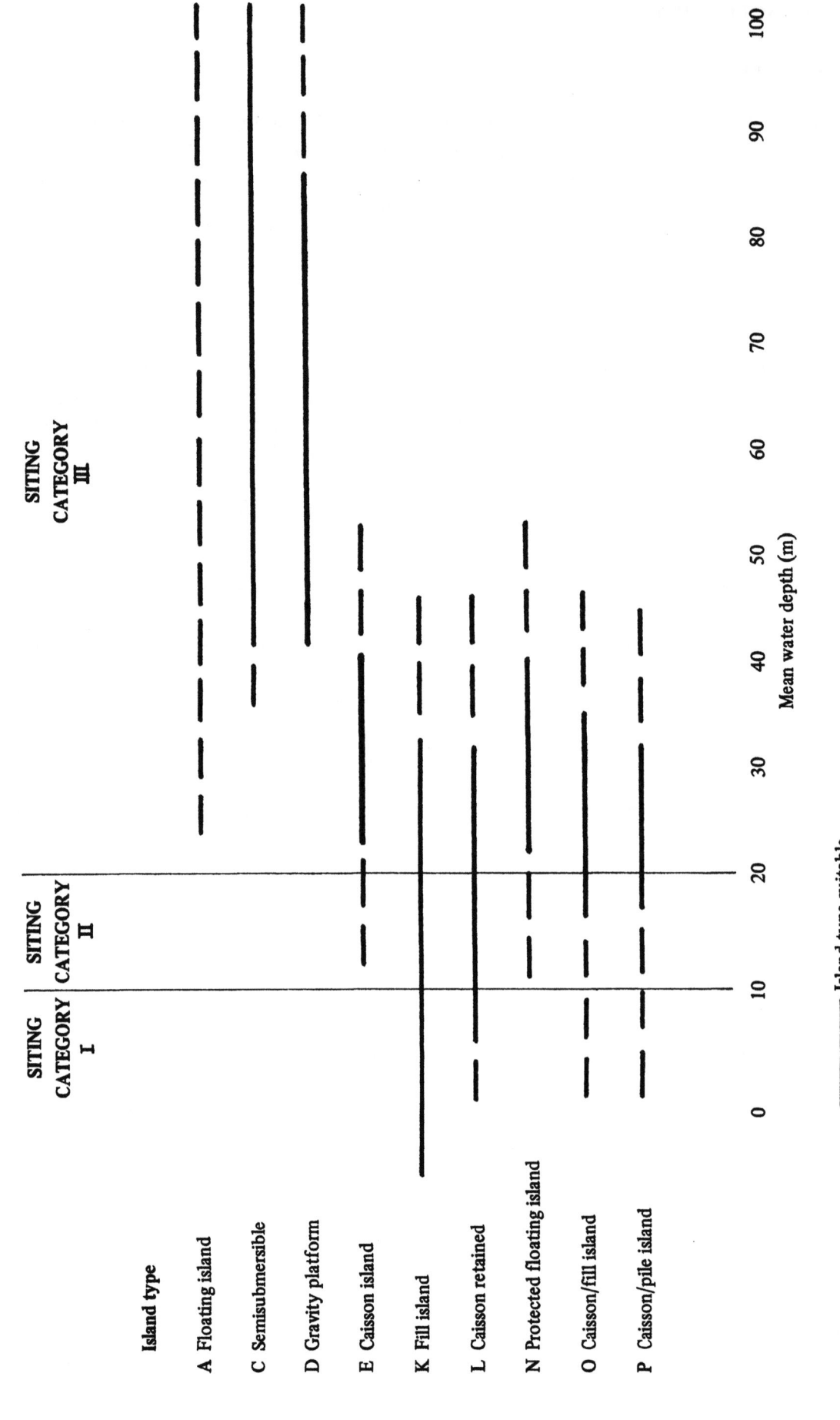

G	Caisson island	— tow out
K	Fill island	— no major difficulties
L	Caisson retained island	— no major difficulties but a fill island is better except possibly in very exposed locations
N	Protected floating island	— breakwater expensive in the deeper water required for a floating island
		— tow out
P	Caisson/fill island	— no major difficulties
Q	Caisson/piled island	— piling limits the suitability to shallow water.

Siting categories - Island suitability

General

11.4   The types of island that could be used in each of the three siting categories are:

Category	Water depth (m)	Suitable types of island (in order of suitability)
I	10 (maximum)	1) fill island 2) caisson retained island
II	10-20 m	1) fill island 2) caisson retained island 3) caisson/fill island 4) protected floating island
III	20 (minimum)	1) fill island (water depth less than (40 m) 2) caisson retained island 3) caisson, caisson/fill or protected floating island depending on exposure 4) semisubmersible (water depth greater than 50 m)

Category I

11.5   The fill island is the most suitable under all conditions in category I. The only other type of island of comparable cost for this siting category is the caisson retained island. The composite caisson/piled and caisson fill islands are much more expensive and will require dredged approaches to bring in the caissons. Their advantages lie in the prefabrication of the power plant on the caissons. These advantages are unlikely to make them more attractive than fill islands in this category.

Category II

11.6   The fill island is the most suitable in category II. Caisson retained islands become increasingly expensive in deeper water. The caisson island, the protected floating island, the caisson/fill island and the caisson/pile are more attractive than in category A but dredging would still be required for these types.

Category III

11.7   The fill island is the most suitable at the shallower depths of category III. Caisson/retained and protected floating islands are more expensive. However, for water depths in excess of 40 m - 50 m not only do costs of fill islands rise disproportionately but also their feasibility is in doubt. More research and development would be required before a fill island could be recommended for deeper waters. The same comments apply to protected floating islands. As water depth increases gravity platforms become more suitable than caisson islands. In very deep water the semisubmersible is the only possible type. The cost of islands suitable for this category is high.

11.8   The suitability of the types of island for various depths of water is summarized in Table 10. Island costs are shown on Drawings 26, 27 and 28.

Exposure/distance offshore

11.9   The degree of exposure possible within each category although affecting the cost of islands is unlikely to alter the choice of the most suitable island type.

11.10   The distance offshore will affect the cost of access and transmission links but again is unlikely to affect the choice of the most suitable island type.

11.11   The overriding factor influencing the choice of island type is water depth.

Timescale

11.12   Power station foundation construction could start on a Category I or II fill island within 2½ years of the award of the fill island contract. A three and a half year lead in, before award, for site investigation, site selection, model tests and design could be expected.

Production/construction facilities

11.13   Fill islands require the least shore facilities. Islands using caissons, floating islands, semisubersibles and gravity platforms require extensive shore facilities. For the latter two types the feasibility of finding sites for suitable construction basins is in doubt. The large quantities of materials required for some types of island, such as rock for fill islands, may not be available locally and may therefore have to be imported from outside the EEC area.

Factors affecting overall costs

11.14   The overall cost of a nuclear power station on an island is dependent not only on the cost of the island but also on the increased cost of constructing the power plant itself, the cost of access to the island and the cost of the transmission link and the cost of auxiliary structures. The first of these factors could involve costs of several times that of the island. The costs, which will be site specific, will be least for inshore fill islands and greatest for gravity platforms.

11.15   The provision of a fixed access will help to reduce power plant construction costs. Although some provision for both sea and air transport will have to be made, they are not as reliable as a fixed access. A fixed access should be provided if at all possible.

11.16   The cost of the transmission link to the shore is directly proportional to the distance offshore. The cost is high. These factors heavily favour an inshore site.

Power station weight and layout

11.17   The areas and weights of the power plant given in our terms of reference are fairly generous. A reduction in the design weight of the power plant would make floating, caisson and composite islands more attractive.

Summary

11.18   The construction of an offshore island to support a nuclear power station is feasible. For the recommended fill type of island the actual design and construction would use proven methods that are well within existing technology.

12. RECOMMENDATIONS

12.1   This study has not been site specific and being within broad terms of reference has, of necessity, been fairly general. The study has identified two specific areas in which further work on offshore islands would, we believe, be justified.

12.2   Firstly, we suggest that the following matters, not strictly within our terms of reference, should be looked at:

　　　　1.   the increased cost of power plants on islands

　　　　2.   the reduction in overall weight and area of the power plant

　　　　3.   in conjunction with (2) the study of small mobile stations

　　　　4.   the integration of the power plant into the island structure.

12.3   Secondly, to make the study useful, we suggest that further work should be carried out within much narrower terms of reference possibly limiting site locations to not more than 3 or 4 specific sites or areas in less than 15 m of water. This second study could:

　　　　1.   identify site specific design criteria

　　　　2.   check choice of type of island

　　　　3.   work up the outline designs to "feasibility" stage

　　　　4.   identify sources for all materials

　　　　5.   confirm costing.

12.4   The first of these studies has direct application to nuclear power station developments in any of the EEC countries. The second study could take place in two parts, items 1 and 2 being of a general nature whilst 3, 4 and possibly 5 would not be required unless and until a potential island development was identified.

APPENDIX A

TECHNICAL ANNEX

Title and subject of the study

Islands for offshore nuclear power stations.
Study on the technologies that could be used for the construction of islands for offshore nuclear power stations in different siting conditions.

Scope of study

The scope of the study will be as follows :

1. To review alternative methods of island construction and its application to nuclear power station siting.

2. To relate alternative methods of construction to three basic categories of siting outlined herewith and to report on the alternatives that are available for each category.

3. To provide a broad based economic comparison of the alternatives in each category for different plant capacities.

4. To provide the timescale for construction of each alternative.

5. To highlight those factors such as material availability and production/construction facilities that could influence the preferred alternative in each category and the economic implications.

Study schedule

1) **Siting categories**

It is suggested that the study should be based upon typical offshore locations belonging to three basic siting categories as follows :

a) **Sites in Sheltered Waters Typified by Estuarial Conditions**
   These sites may be located in the inter-tidal zone but would be limited to overall water depths of 10 metres and be situated not more than 4 kilometers from high water mark.

b) **Sites Offshore From the Main Coastline but in Depths of Water Not Exceeding 20 metres.**
   These sites should be considered to be within 5 kilometres of high water mark.

c) **Sites Offshore From the Main Coastline in Exposed Condisions Where the Water Depths Exceed 20 metres**
   These sites should be not less than 5 kilometres from high water mark.

./.

2) **Type of Plant**

The types of nuclear plant to be considered should be of the water and gas cooled types.

Plant capacities should be based on station developments of 2500 MW but should consider total generating capacities of 5000 and 10000 MW.

The Commission will make arrangements with the members of the working group offshore nuclear power plants in order to provide the necessary informations for the contractor, on the following matters :

- the sizes, weights and layout of the main items of plant and buildings

- the short term movements and long term settlements that the elements of a power station could tolerate

- outline cooling water arrangements

- methods of transmitting electricity ashore

- the marine and/or land-based plant that would be required for maintenance.

APPENDIX B
EEC - Islands for Offshore Power Stations
Basic Parameters for Guidance of the Consultants

1. Introduction

The Terms of Reference for the Study state that both water and gas cooled reactor types should be considered. However because of the greater physical size and ground loadings of a gas cooled reactor compared to a water cooled reactor, it is considered that a gas cooled reactor would be unsuitable for a floating or fixed platform type of island construction. Consequently, for the purpose of this study the date appliable to a water reactor should be used, although for comparative purposes, data for both types are included.

2. Plant Types and Capacities

For this study, artificial islands/platforms suitable for a station development of 2500 MW in units of 1300 MW are to be considered but with the possibility of extension to 5000 MW and 10,000 MW. For a water cooled reactor station, each 1300 MW unit will consist of a single 1300 MW reactor and turbine; for a gas cooled reactor station each 1300 MW unit will consist of 2 x 650 MW reactors and two turbines.

3. Main Plant Layouts and Ground Loadings

The buildings for the main plant items for each 1300 MW unit (reactor building and auxiliary services, and turbine hall) can be accommodated within a rectangular area of about 250 m x 200 m. Within this area, the maximum ground loading is 650 KN/m^2 for a gas cooled reactor, and 450 KN/m^2 for a water cooled reactor. Typical layouts of main plant for both a water cooled reactor and a gas cooled reactor, showing the ground loading of individual buildings, are attached - Figs. 1 and 2.

4. Auxiliary Plant Buildings

In addition to the main plant building, additional space must be allocated for auxiliary plant items (CW pumphouse, switchgear etc.) and workshops, stores, and administrative buildings. Taking account of this, the total area required for each unit of 1300 MW is estimated to be about 15 ha. This is based on the experience of land-based sites. In situations where the site area is for economic reasons, a critical feature, (as the case for floating platforms) this area could be reduced considerably, possibly by constructing multiple storey platforms. For this situation, the study should assume a platform of 5 ha. for the main reactor building for a 1300 MW unit (as specified in Section 3) and should consider the implications of a range of platform sizes in multiples of 5 ha. up to 15 ha. to accommodate the auxiliary plant.

5. Construction Area

Typically, for land based sites, construction areas are in the range 20 - 30 ha. However for offshore sites the need for a construction area will depend largely on the extent of prefabrication that is possible on the mainland. In the case of floating platforms constructed at a mainland dock, the amount of construction area required at the offshore site would be minimal.

6. Additional Requirements

The figures and layouts given above are estimates based on the knowledge of land based stations. However for an offshore station additional features will be required, for example.

(a) A floating port/harbour to provide access for equipment and materials, and for Personnel transport during construction and subsequent operation.

(b) A heliport for emergency evacuation and Personnel transport in rough weather and for transporting essential supplies.

(c) Accommodation designed for long stays on the island, both during construction and subsequent operation.

(d) A shielded emergency area in case of a serious incident.

7. Labour Force

(a) Gas Cooled Reactor Station.

The labour force for 2 x 1300 MW stations would be about 1000 allowing for shift working. During construction the labour force would peak at about 1500 - 2000 people, based on current land based construction methods.

(b) Water Cooled Reactor Station
(EdF to complete).

Depending on the method of construction and the scope for prefabrication, the number of construction Personnel on an island/platform site could be considerably reduced.

8. Earthquake Effects and Settlements

The main reactor building foundations must be designed to withstand a peak free field acceleration of 25% g with a peak free field velocity of 25/30cm per second. The maximum allowable tilt is 1:1000.

9. CW Requirements

A CW flow rate of 50 m^3/s should be assumed for each 1000 MW of plant. The CW system must be designed to minimise re-circulation.

10. Transportation Requirements during Normal Operation

Provision must be made for the regular transportation from the island/

platform of the irradiated fuel flasks, a gross weight of about 60 tonnes. The largest single component requiring transportation for maintenance purposes would be the generator stator (nett weight about 400 tonnes for a 1300 MW set).

11. Transmission Requirements

Dependent upon the relative location of the island with the power deficits of the grid system, it is expected that up to 5 GW could be connected at 400 kV AC. The offshore connection would consist of four 3 phase 2.5 GW routes of 400 kV single phase cable buried in trenches 1.5-2.0 metres deep.

TABLE B-1 Terminology

		Items	Area /1250 MW
Power station	Main plant	Reactor Turbines Generators Control room Plant auxiliaries	5 ha
	Auxiliary structures	Switchgear Cooling water Administration bldgs. Harbour etc.	10 ha maximum
Ancillaries		Bridge Tunnel Transmission Work harbour on land etc.	

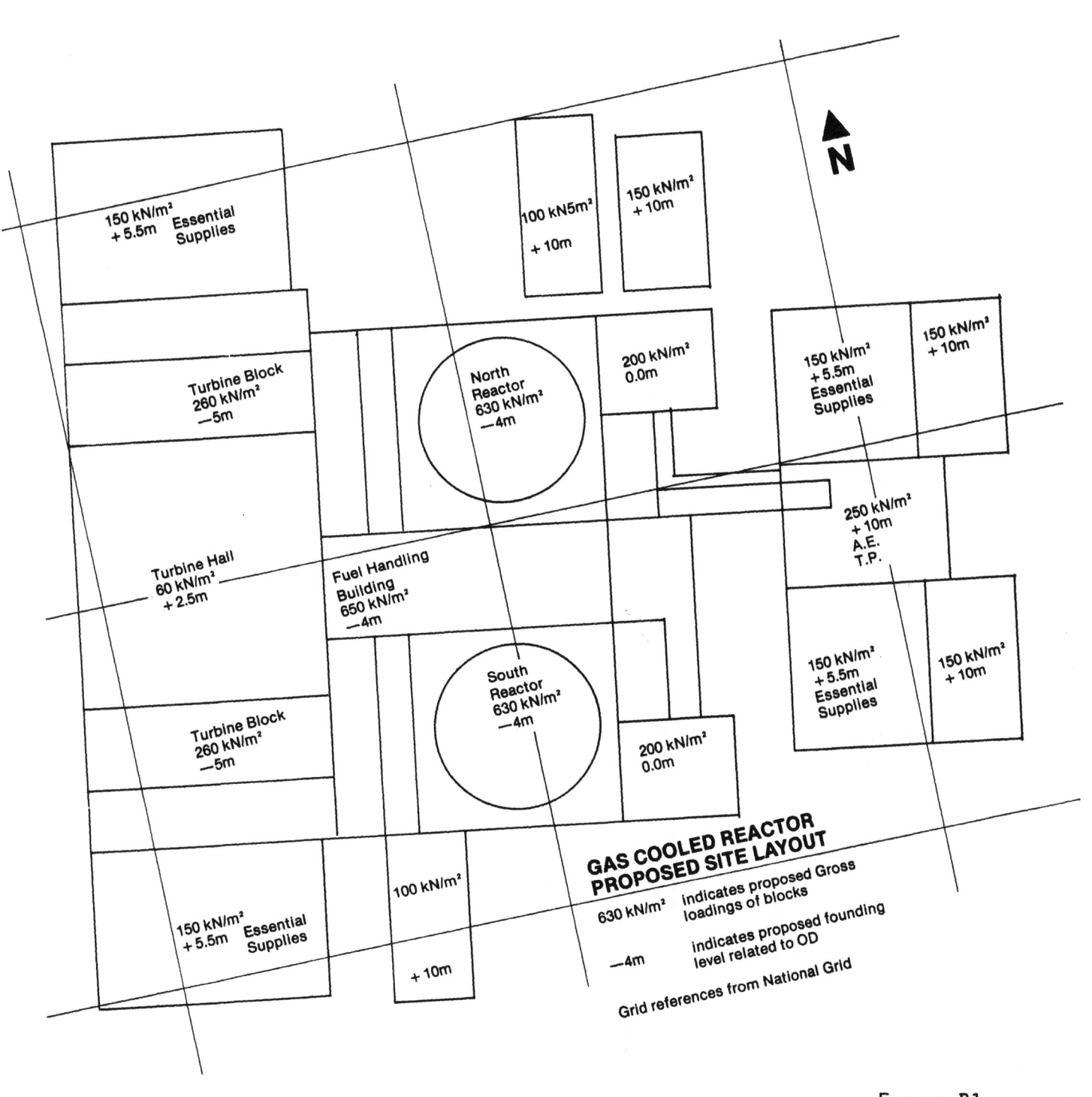

FIGURE B1

WATER COOLED REACTOR

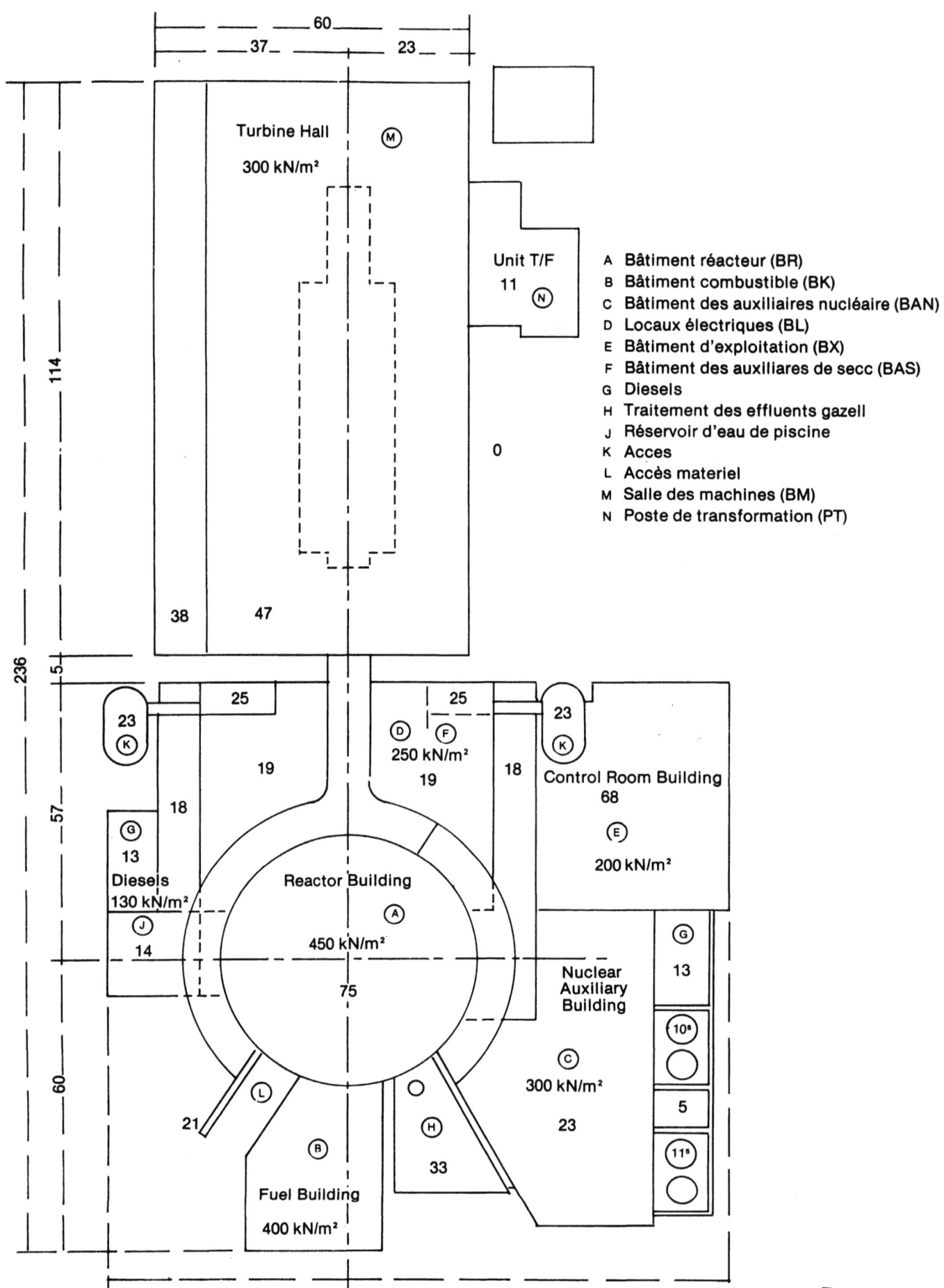

Figure B2

APPENDIX C

SELECTED BIBLIOGRAPHY

1. Offshore islands

1.1 R.R. Adams. 'Offshore rock island construction'. 6th international harbour congress, Hamburg 1974.

1.2 J.A. Blume. 'Rincon offshore island and open causeway' Journal of the ASCE, Waterways and Harbours Division. September 1959.

1.3 J. Bonasia, M. J. McCarthy, E.J. Schmeltz. 'Ocean Engineering in design of sea islands'. Proceedings of conference on civil engineering in the oceans, Newark 1975.

1.4 J.C. Bruce & J. J. A. de Jong. 'Design and construction of a caisson retained island drilling platform for the Beaufort sea'. Offshore technology conference, Houston 1978.

1.5 'Building a multi-purpose island in the sea'. Ocean industry, April 1973.

1.6 CIRIA, Underwater Engineering Group. 'New applications for concrete offshore'. Draft report 1980.

1.7 Deutsche Babcock. 'Concept for a mobile offshore power station'.

1.8 B.E.W. Dowse. 'The role of hybrid islands for offshore coal mining', Oceanology International 80.

1.9 B.E.W. Dowse. 'Sandisle structures'. Ground Engineering, March 1979.

1.10 A.G.F. Eddie. 'Offshore structures at Hay Point, Queensland'.

1.11 A.G.F. Eddie & U.B. Hansen. 'The utilization of large concrete caissons in various sea conditions'. 3rd Australian conference on coastal & ocean engineering, April 1977.

1.12 Floating islands Group. 'Floating Islands'.

1.13 Fraenkel & Triggs. 'Special features of the civil engineering works at Aberthaw Power Station'. ICE proceeding 1962.

1.14 F.J. Hansen. 'Contruction and installation of concrete ocean caissons'. FIP Symposium on concrete sea structures, 1972.

1.15 E.H. Harlow. 'Breakwater and mooring system for floating nuclear power plants'.

1.16 E.H. Harlow. 'Offshore floating terminals'. Journal of the ASCE Waterways, Port, Coastal & Ocean Division. August 1971.

1.17 E.H. Harlow. 'Offshore islands'. Journal of the ASCE Waterways, Port, Coastal & Ocean Division. February 1977.

1.18 E.H. Harlow & M. Kehnemuyi. 'Mooring system for Atlantic generating station.' Offshore technology conference, May 1974.

1.19 E.H. Harlow & M. Kehnemuyi. 'Breakwater for Atlantic generating station' Offshore technology conference, May 1974.

1.20 J.M. Jordaan. 'Artificial Island Engineering'. Proceedings of conference on civil engineering in the oceans, Newark 1975.

1.21 M. Kehnemuyi. 'Floating nuclear electric generating plants'.

1.22 R.E. Lochbaum. 'The concept and design of floating nuclear power plant for Atlantic stations No. 1 & 2'.

1.23 'Man-Made Island in Tokyo Bay'. International Construction, September 1974.

1.24 'Offshore Structures'. New Civil Engineer, Supplement 1974.

1.25 N.A. Smith. 'Sea towers of sand'. HK Engineer, July 1978.

1.26 P. Soros. 'Artificial Islands for offshore transshipping terminals'. 6th International Harbour Congress.

1.27 P.L. Stuart. 'BP West Sole gas platform structures in the southern North sea'. Conference on maintenance of maritime structures. October 1977.

1.28 R.J. Tatge, E.T. Hillberg & B. J. Washom. 'The evaluation of proposed offshore LNG receiving terminals'. Offshore Technology Conference, Houston 1978.

1.29 M. Xercavins. 'Offshore oil storage in the North sea-Ekofisk reservoir'. FIP Symposium on concrete sea structures 1972.

Note: Reports of studies carried out in Belgium, Germany and Holland have also been made available to us. The sources of these documents are acknowledged in Appendix D.

2. **Background references**

2.1 E.E. Adams, C. W. Almquist, D.V. Ingraham, K. D. Stolzenbach. Waste heat discharges from offshore power plants. Proceedings of conference on civil engineering in the oceans, Newark 1975.

2.2 P. Ackers. 'Modelling of heated water discharges'. Proceedings of National symposium on thermal pollution, Nashville, Tennessee, 1968.

2.3 P. Ackers, J.D. Pitt, G. Thompson, K.G. Rippin. 'Dispersion of cooling water from a coastal LNG plant'. Proceedings of 17th International Conference on Coastal Engineering, Sydney, 1980.

2.4 R.T.L. Allen & J. Gregory-Cullen. 'Inspection, maintenance and repair of concrete offshore structure'. Cement and Concrete Association, 1974.

2.5 H.C. Berkin & W.F. Trappen. 'A capital cost comparison of floating and land based nuclear plants utilizing risk analysis'.

2.6 G.J. Blight & R.Y.T. Dai. 'Resistance of offshore barges and required tug horsepower'. Offshore Technology Conference, Houston 1978.

2.7 British Standards Institute. 'Code of Practice on Maritime Structures'. Draft for public comment, 1979.

2.8 CIRIA, report 61. 'Design of rip rap slope protection against wind waves'.

2.9 CIRIA, Technical Note 84. 'Rip rap design for wind – wave attack; prototype tests on the offshore bank in the Wash'.

2.10 CIRIA, Underwater Engineering Group. 'New applications for concrete offshore'. Draft Report, February 1980.

2.11 Department of Energy. 'Guidance on the design and construction of offshore installations'.

2.12 Det Norske Veritas. 'Rules for the design, construction and inspection of offshore structures'.

2.13 European Federation of corrosion. Proceedings of two-day joint conference on long service from offshore structures. February 1976.

2.14 Federation internationale De La Preconstrainte (FIP). 'Recommendations for design and construction of concrete sea structures'.

2.15 J.A. Fischer, I Watson, H. Singh, T.D. Lu, S. Saxena, D. Koutsoftas, L. Stahl. 'Design forces for offshore nuclear power plant consrtuction'. Proceedings of conference on civil engineering in the oceans, Newark 1975.

2.16 F.J. Hansen. 'Concrete construction offshore' Oceanology International '69.

2.17 H.G. Herrmann and W.N. Houston. 'Behaviour of seafloor soils subjected to cyclic loading'. Offshore Technology Conference, Houston 1978.

2.18 Hogben, Miller, Searle and Ward. 'Estimation of fluid loading on offshore structures'. National Maritime Institute, April 1977.

2.19 N. Hogben & R. G. Standing. 'Wave loads on large bodies' NPL Conference, April 1974.

2.20 R.Y. Hudson. 'Concrete armour units for protection against wave attack'. Misc Paper 11-74-2 US Army Engineer Waterways Experiment station.

2.21 Institution of Civil Engineers. 'The Civil Engineer in War'. Articles on Mulberry harbours and other wartime offshore structures.

2.22 Institution of Civil Engineers. 'Proceedings of conference on design and construction of offshore structures'. London, October 1976.

2.23 P. Kaplan, T. P. Sargent. 'Motions of offshore structures as influenced by moorings and positioning systems'. Proceedings of conference on behaviour of offshore structures, 1976.

2.24 National Maritime Institute. 'The risk of ship/platform encounters in UK waters'. Report May 1978.

2.25 J. M. Niedzwecki. 'A comparison of non-metallic ropes with wire rope and chain mooring lines for deep water application'. Offshore Technology Conference, Houston 1978.

2.26 G. Somerville & H. P. J. Taylor. 'Concrete in the Oceans'. Cement and Concrete Association Report 1974.

2.27 US Army Coastal Engineering Research Centre. 'Shore Protection Manual'.

2.28 R.L.P. Verley. 'Wave forces on structures an introduction'. BHRA technical note 1319.

2.29 B.J. Watt et al. 'Earthquake survivability of concrete platforms'. Offshore Technology Conference, Houston 1978.

2.30 P.J. Wemelsfelder. 'On the use of frequency curves of stormfloods'. 7th Conference on coastal engineering, The Hague, August 1960.

APPENDIX D

WATER DEPTHS WITHIN STUDY AREA

D1 Water depths in the Western Approaches, the North Sea and the Mediterranean are shown on Drawings D1, D2 and D3 respectively. Because the contours for 10 m and 20 m depth generally occur too close to the coast to be shown clearly an alternative way of presenting the data has been chosen. The coastal waters have been classified according to the six divisions shown on Table D1. These divisions are represented on the maps by different markings along the coast.

Table D1 — Divisions of Coastal Waters

Depth	Distance from shore to this depth
10 m	Less than 5 km
10 m	Between 5 and 20 km
10 m	More than 20 km
20 m	Less than 5 km
20 m	Between 5 and 20 km
20 m	More than 20 km

D.2 Depth contours for 200 m and 1000 m are also shown on the maps.

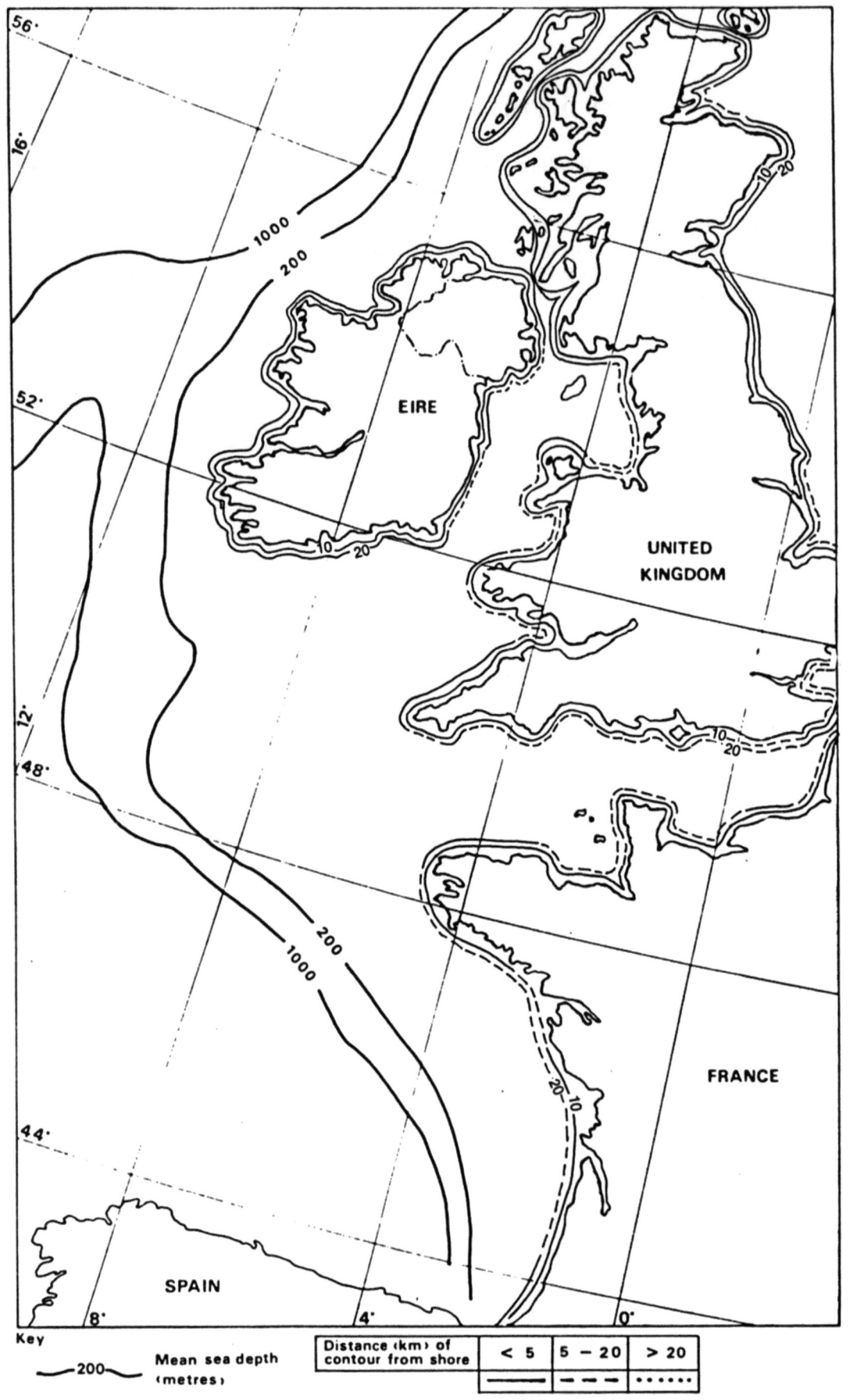

Western approaches - Mean sea depths Drawing D1

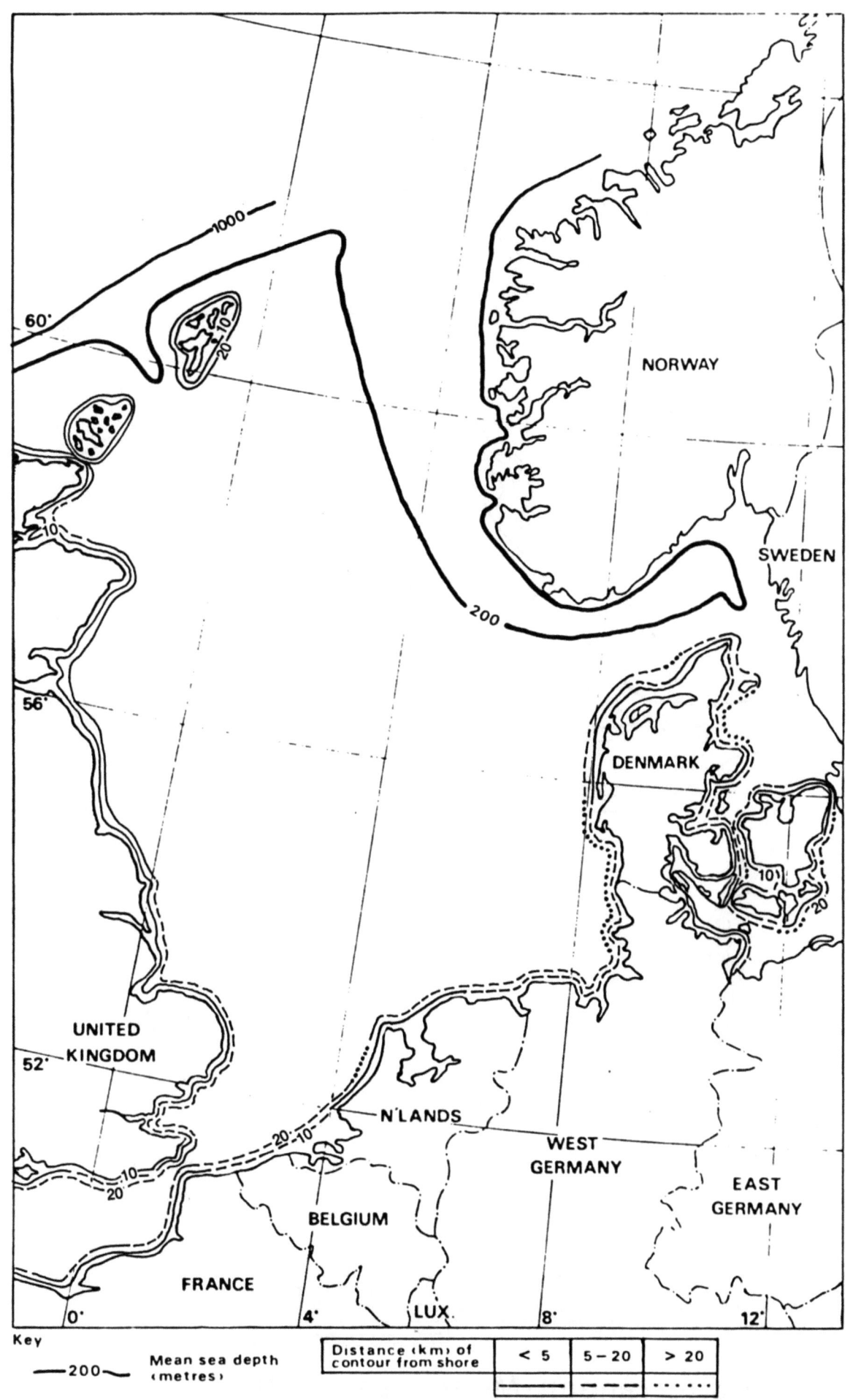

North sea - Mean sea depths　　Drawing D2

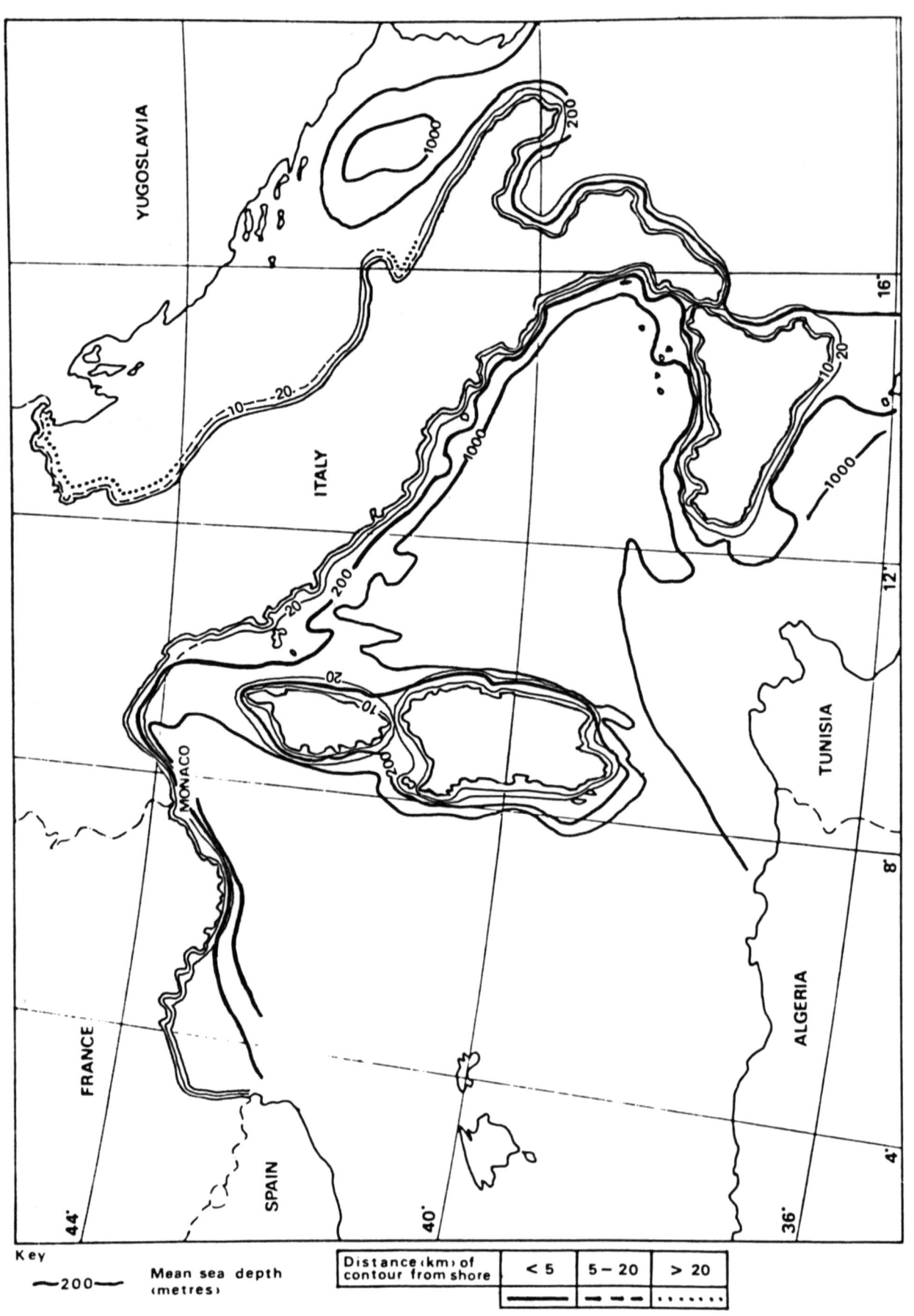

Mediterranean-Mean sea depths　　Drawing D3

APPENDIX E

WAVE CLIMATES

E1. Offshore wave climate

E1.1 Data on wave climates offshore are available in two forms - as records from instruments (ship borne wave recorders, wave rider buoys etc) or as visual observations taken from ships in passage. The former source is far more reliable and accurate but data have only been recorded in limited areas. The North Sea and UK coastal waters have the most comprehensive data sets and these have been analysed by Draper (1972) and can be presented in the form of contours showing the value of significant wave height having a return period of 50 years (H_{50}) Drawings E1 and E2 reproduce this data on wave height. Visual observations of wave height for Mediterranean areas and for the West coast of France have been obtained from the U.S. Naval Weather Service Command (1974) and have been analysed to produce the results shown in drawings E2 and E3.

E1.2 So far as offshore islands are concerned, it is clearly inadequate to design on the basis of a return period of 50 years as there is a high probability that this event will occur in the lifetime of the structure. By introducing the concept of "encounter probability" a more rational design wave height can be evaluated. Table E1 indicates the factor by which values from drawings E1, E2 and E3 should be multiplied to give design wave heights (for various return periods). The principal limitation of this approach is the need to extrapolate from short periods of record (2-10 years long) to long return periods. A much more rigorous analysis is required at design stage.

E2. Nearshore wave climate

E2. In the case of islands constructed in shallow coastal waters, the wave data for offshore areas cannot be used directly. The affects of refraction, friction, shoaling, diffraction and breaking must be taken into account. Whilst there are accepted methods for evaluating these effects, it is rarely possible to make accurate forecasts of the likely inshore wave climate on the basis of offshore data alone and, in practice, wave records for the specific site considered must be obtained.

TABLE E1: Approximate relationship between design wave height H_D (R yrs) and H_{50} and encounter probability in 100 and 1000 years.

Return period R	Probability of H_D (R yrs) Occurring in 100 yrs	Probability of H_D (R yrs) Occurring in 1000 yrs	Approx value of $\dfrac{H_D \text{ (R yrs)}}{H_{50}}$
10	99.99%	99.99%	0.89
50	86%	99.99%	1.00
250	33%	98.2%	1.05
1 000	9.5%	63.2%	1.23
5 000	2.0%	18.1%	1.35
25 000	0.4%	3.9%	1.48
100 000	0.1%	1.0%	1.58

Example: If design life of structure is 100 years and the probability of the design wave being exceeded is 0.1% then $H_D = 1.58\ H_{50}$

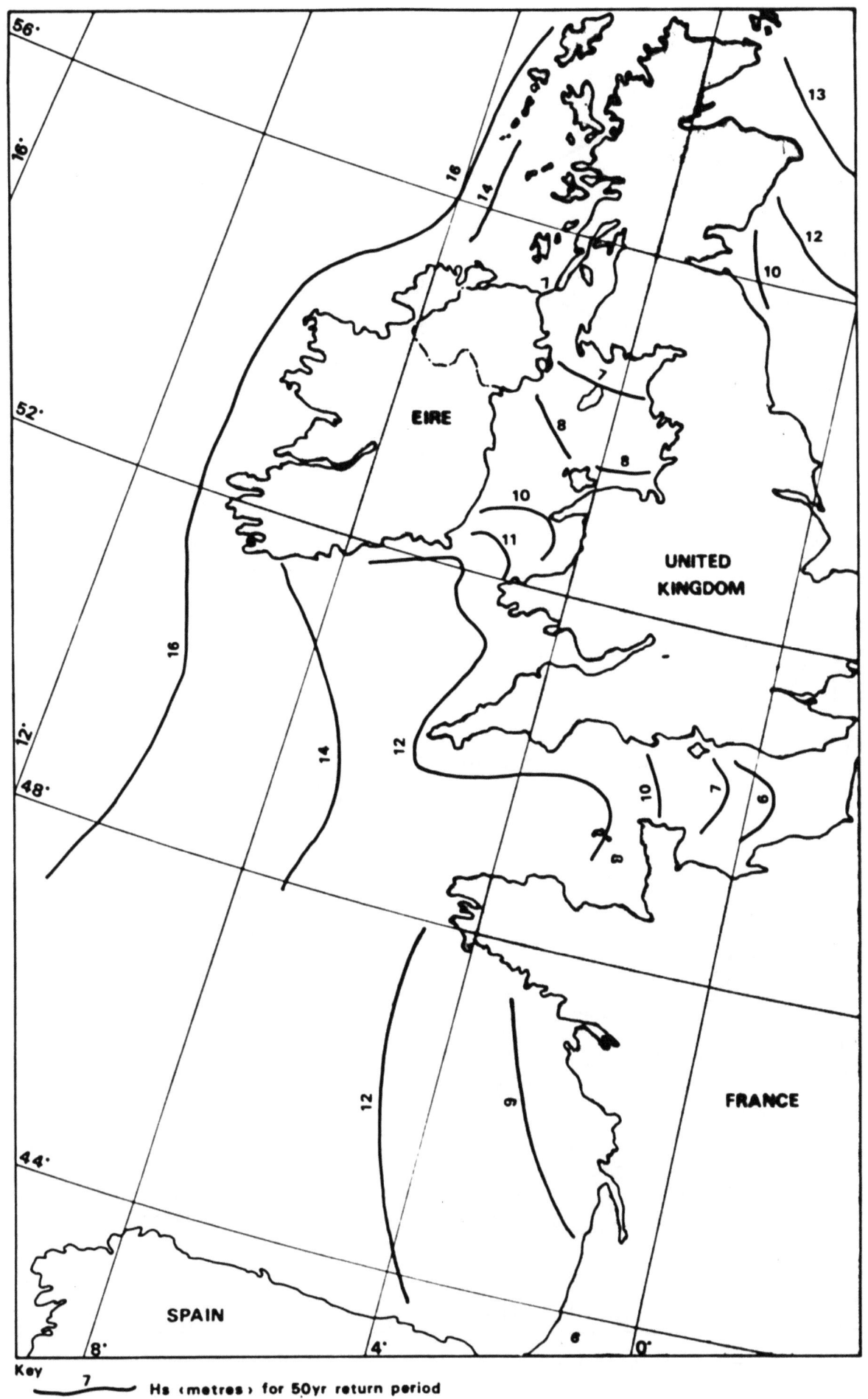

Western approaches - Wave heights Drawing E1

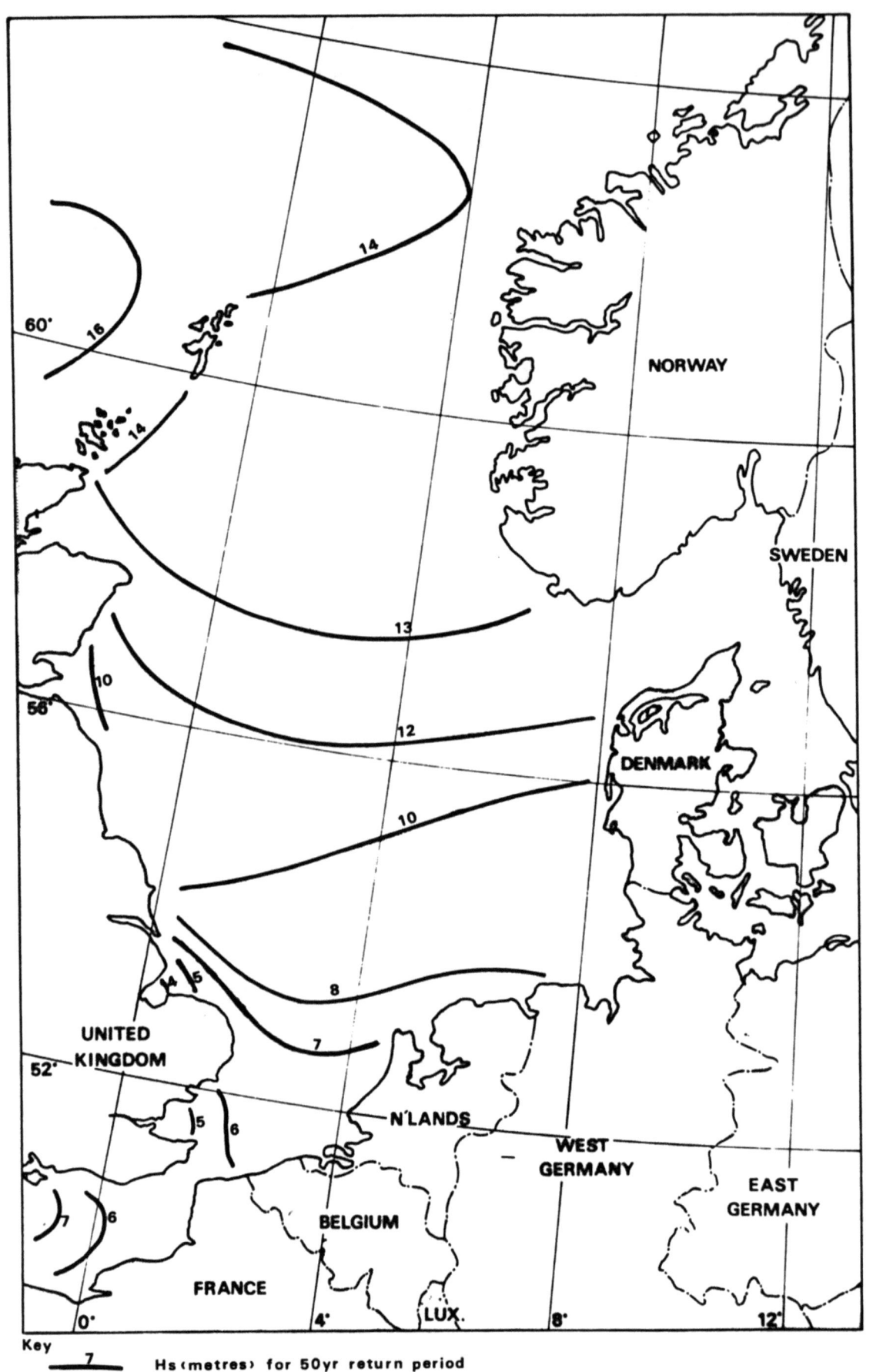

North sea - Wave heights Drawing E2

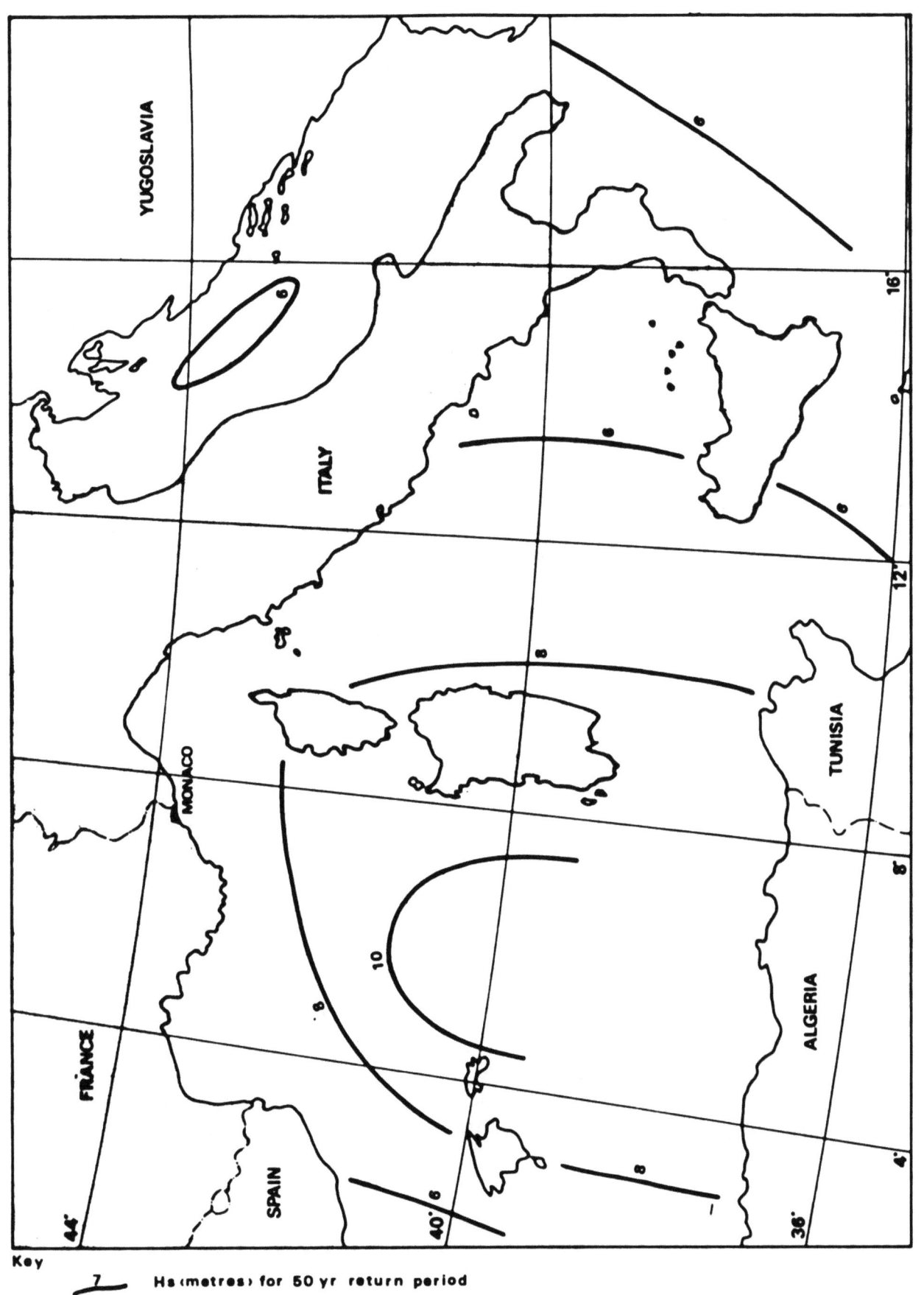

Mediterranean - Wave heights Drawing E3

APPENDIX F

WAVE FORCES ON STRUCTURES

F1. Introduction

F1.1 Forces on a wide range of structures differing in both shape and size need to be evaluated for a wide range of wave conditions. We have sought to identify the most critical condition for each structure with sufficient precision to make a reasonable estimate of the main dimensions and hence cost of the structure. To do this we have first classified the flow regime according to the structure's size and the wave length of the incident waves. We have then used a method appropriate to this flow regime to estimate the magnitude of forces and moments due to wave action.

F2. Regimes of flow

F2.1 The method used to calculate forces is governed principally by the size of the structure and the design wave parameters. The first step is to establish the regime of flow and hence the appropriate method. The dominant parameter is the ratio D/L (diameter of structure/wave length) and the main classifications are:

Regime		Method of determining forces	Typical application
(i)	$D/L < 0.2$	Morison's equation	slender structures such as piles where wave kinematics are not significantly affected by the structure.
(ii)	$1.0 > D/L > 0.2$	Diffraction theory	wave forces on large diameter structures
(iii)	$D/L > 1.0$	Reflection	waves on long walls, breakwaters, etc.

When using Morison's equation a further subdivision may be made depending on the ratio D/H (diameter of structure/wave height). For values of D/H greater than 1 drag can be neglected. Figure G1 shows the different regimes in relation to D/L and D/H together with points representing some typical structures. The methods of evaluating forces for each of the flow regimes are described below.

F3. Morison's Equation

F3.1 Morison's equation is the commonly used method of estimating forces on slender structures such as piles. There are two components, a drag force and an inertia force, both of which involve a semi-empirical coefficient. These coefficients vary according to flow conditions, surface roughness and other factors. The full form of Morison's equation is:—

$$F = \tfrac{1}{2} C_d \rho D U |U| \cos\theta + \frac{C_m \pi D^2 \dot{U}}{4} \sin\theta$$

$$\text{drag} \text{inertia}$$

Where   F   is the force per unit length of pile
D   is the pile diameter
u   is the water particle velocity and
$\dot{u}$ is the water particle acceleration
Cd   is the drag coefficient, generally in the range 0.6–1.5
Cm   is the inertia coefficient, generally in the range 1.5–2.0
ρ is water density
θ is the phase angle of the wave relative to the structure and is given by $\theta = \frac{2\pi}{T} \cdot t$: t being time

To evaluate the total force on a pile the force F has to be summed by integration over the length of the pile.

F3.2 The difficulties in applying the equation are twofold:—

(i) the uncertainties in allocating values to Cd and Cm

(ii) the complexities of wave kinematics which are needed to evaluate u and $\dot{u}$.

Guidance in selecting Cd and Cm is given in the literature but for preliminary estimates values of Cd = 2 and Cm = 1 are used. Use of linear wave theories simplifies claculation of u and u but the procedure is still lengthy. A computer program has been written to evaluate forces and moments from Morrison's equation using linear theory. It should be noted that this theory is not strictly valid for shallow water or large waves.

F3.3 If the ratio of the structure diameter to wave height is greater than about 1 the drag component becomes negligible in comparison with the inertia component. In the case of a vertical cyclinder extending through the full water depth and when inertia is the dominant force the maximum force is given by the equation:

$$F_{max} = \frac{C_m \rho \pi^2 D^2 L H}{4 T^2}$$

F4. Diffraction

F4.1 One of the basic conditions of applying Morison's equation is that the structure is small in relation to wave length so that the structure has negligible effect on the water particle movement. If the ratio D/L exceeds about 0.2 then this is no longer the case. The velocity potential $\emptyset$ comprises an incident potential $\emptyset_i$ plus a scatter potential $\emptyset_s$. $\emptyset_i$ is defined by the wave motion only but $\emptyset_s$ depends on the structure. The calculation of $\emptyset_s$ and subsequent calculation of pressure, velocities and accelerations is called diffraction analysis. In general the solution can only be determined by numerical methods but for simple geometries there are analytical solutions.

F4.2 The Beach Erosion Board, Technical Memo No. 69, 1954, 'Wave forces on piles, a diffraction theory' gives a method for calculating forces on vertical cylinders extending through the whole water depth. The method has been shown to be equivalent to using Morison's equation with:

$$C_d = 0$$

$$\text{and } C_m = \frac{4 L^2 f_A}{\pi^3 D^2}$$

where f_A is a complex function of π D/L. Figure G2 shows the variation of C_m with π D/L. Putting $C_d = 0$ and taking a suitable value for C_m, the program for evaluating forces and moments from Morison's equation can be used to calculate diffraction forces on a cylinder extending to the seabed. In the case of cylinders extending only part way, i.e. floating structures, the program can be used to give a first estimate of the force.

F5. Reflection

Non-breaking waves

F5.1 Of the several theories available Sainflou's method has been used for this study. Although this can overestimate wave forces it avoids the complexities of other methods. Figure F3 shows the pressure distribution and equations involved in Sainflou's method.

Breaking waves

F5.2 A number of methods for calculating forces from breaking waves have been developed and differ considerably in the forces they predict. The commonly quoted 'Minikin' formula is based on experiments involving trapped pockets of air being compressed and exploding. Such a severe situation cannot be envisaged, except locally, for structures on a relatively flat seabed and in deep water. After a review of the other methods available a method based on that described in the National Maritime Institute, Report R11, April 1977, entitled 'Estimation of fluid loading on offshore structures', has been used. Details of this method are shown in Figure F4.

Broken waves

F5.3   Calculations of forces from broken waves are imprecise and generally the forces are much lower than for the two wave conditions considered above. Since non-broken and breaking waves are dominant in determining the design, methods of calculating the forces from broken waves have not been compared. However, methods of inducing waves to break and the action of broken waves on beach type defences need to be considered when comparing shore protection systems.

F6.   Limitations

F6.1   The calculation methods presented here use linear wave theory although this is only strictly valid for small waves and deep water. Errors increase as the water depth decreases or wave height increases. For example using linear wave theory to evaluate the forces due to a 12 m, 10 s wave on a caisson in 20 m depth of water, the error will be an underestimate of about 10% compared with more sophisticated approaches.

F6.2   The methods proposed for evaluating wave forces by Morison's equation or using diffraction theory do not take account of phenomena such as wave slamming or vortex shedding. For the massive structures being considered in this study such forces should be of only secondary importance to the overall stability. They may be of much greater significance locally and govern the size of some members and would need to be evaluated carefully for a detailed design.

F6.3   No allowance has been made for drag forces on the base of floating structures or for interaction between sections of complex structures e.g. closely spaced piles.

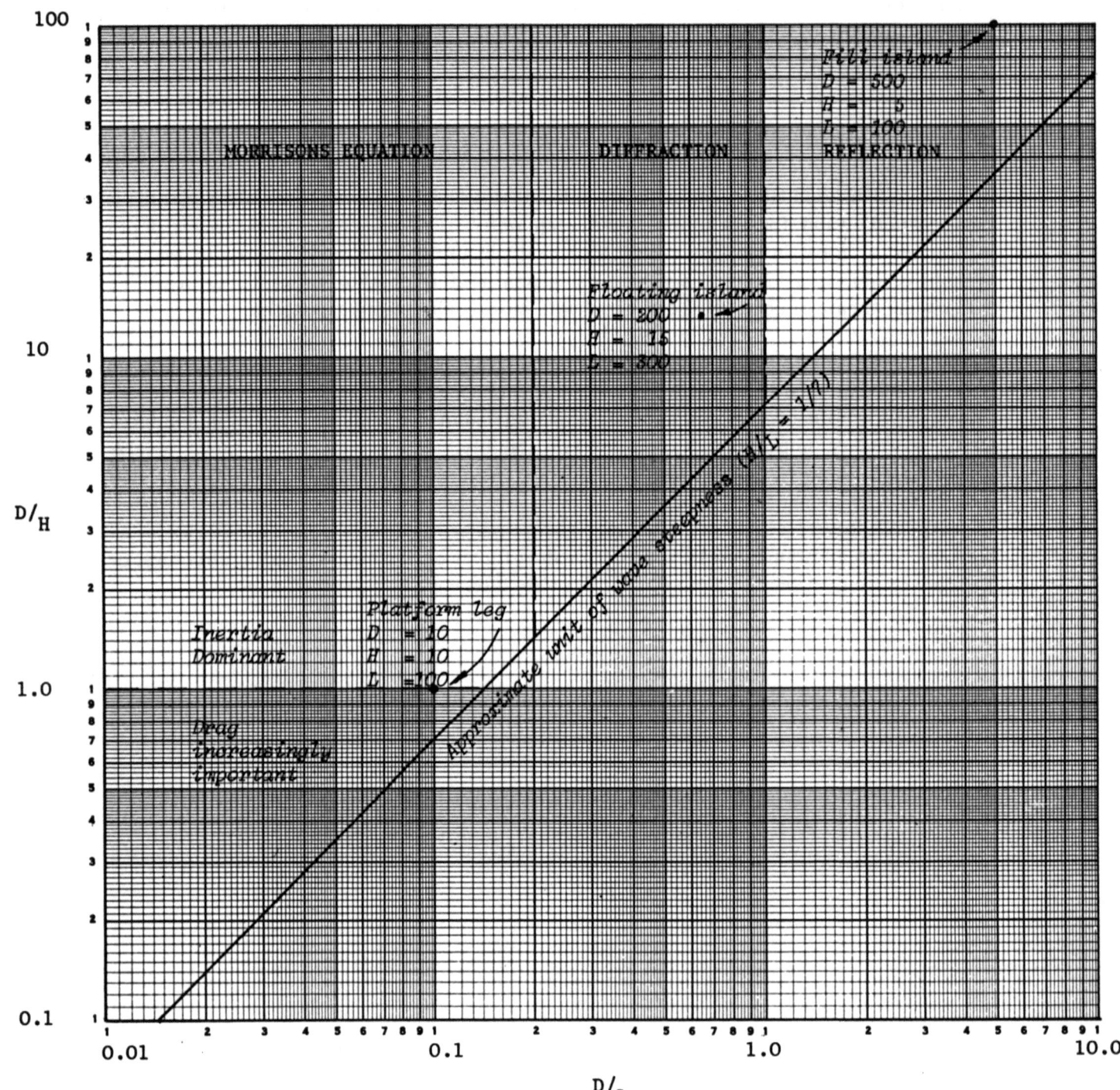

Regimes of Flow

Figure F1

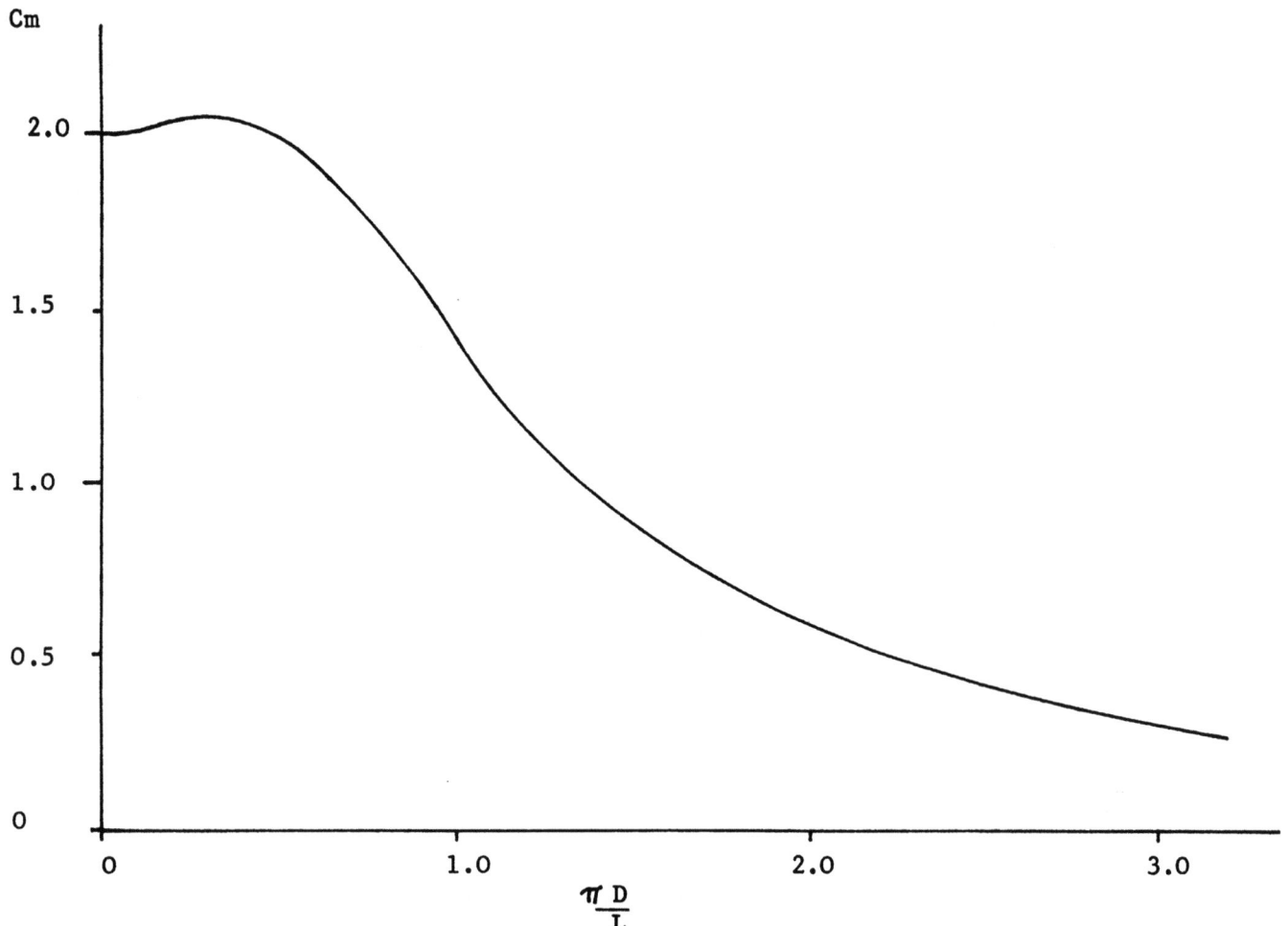

$$C_m = \frac{4F_A}{\pi (\pi D/L)^2}$$

where $F_A = \dfrac{1}{\left\{[J_1'(\frac{\pi D}{L})]^2 + [Y_1'(\frac{\pi D}{L})^2]\right\}^{\frac{1}{2}}}$

and J_1 & Y_1 are Bessel Functions of the first and second kind and $'$ indicates first derivative.

Values of C_m for diffraction analysis

Figure F2

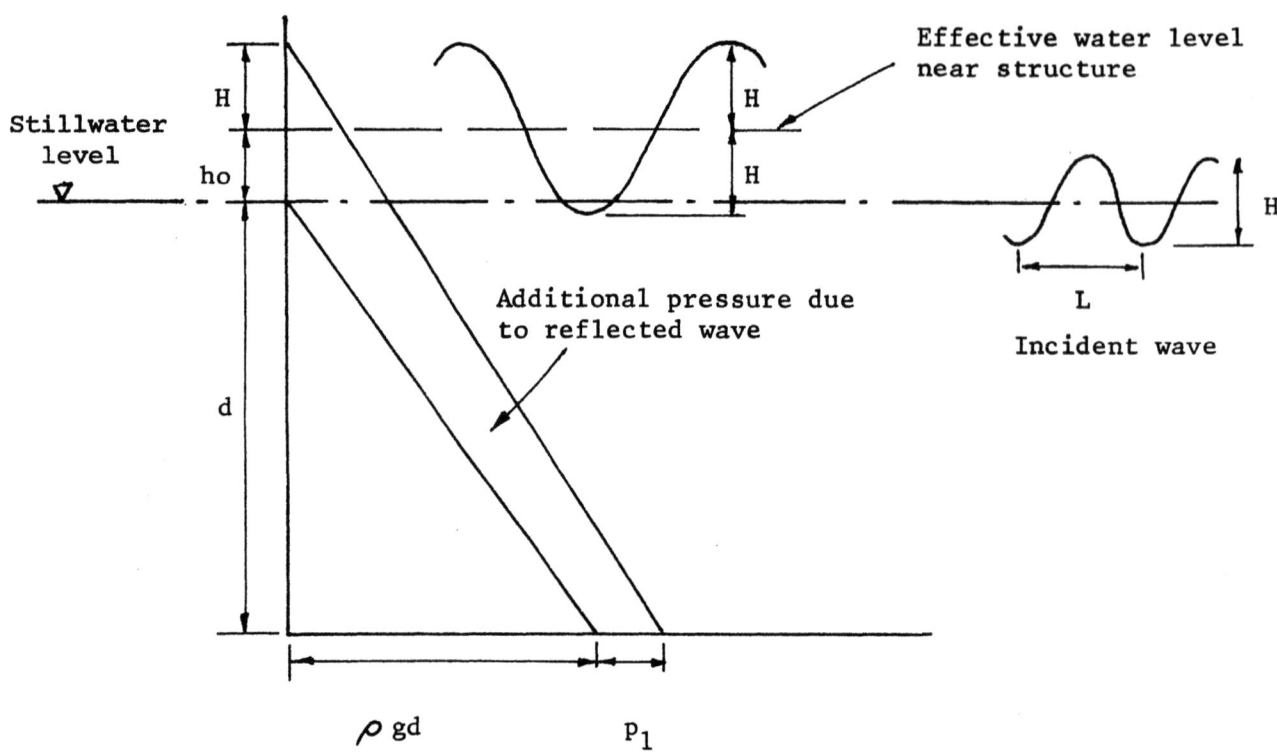

Sainflou's method for calculating forces due to reflected wave:

$$P_1 = \rho g \frac{H}{\cosh(2\pi d/L)}$$

$$h_o = \frac{\pi H^2}{L} \coth(2\pi d/L)$$

Hydrostatic force on structure, without wave:

$$F_H = \tfrac{1}{2}\rho g d^2$$

Total force on structure, including wave:

$$F_T = \tfrac{1}{2}\rho g (d+h_o+H)\left(\frac{p_1}{\rho g} + d\right)$$

Force due to wave:

$$F_W = F_T - F_H$$

Force due to reflected wave

Figure F3

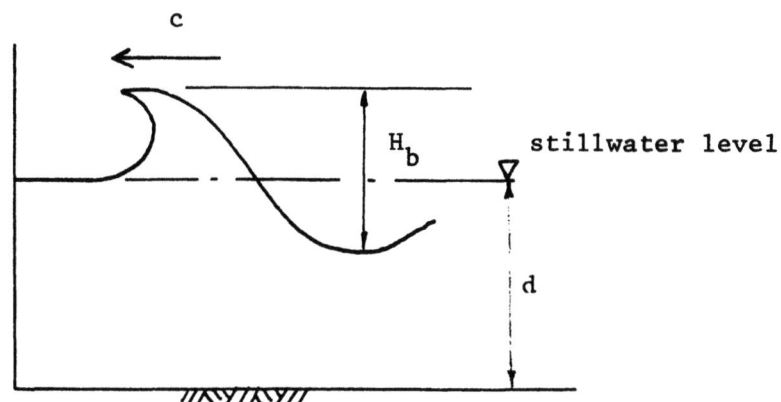

Slamming force $F_s = \frac{1}{2} \rho\, C_s\, H_b\, c^2$

where C_s coefficient in the range 2-6

H_b breaker height

c wave velocity

in shallow water $c = \sqrt{gd}$

and $F_s = \frac{1}{2} \rho\, C_s\, H_b\, g\, d$

Force due to breaking wave

APPENDIX G

TIDAL RANGES WITHIN STUDY AREA

G.1   Tidal ranges in the Western Approaches, the North Sea and the Mediterranean are shown on Drawings G1, G2 and G3 respectively. The information for the Western Approaches and North Sea is plotted in the form of co-tidal lines. Because the variation in tidal range is small in the Mediterranean, ranges at selected points are plotted.

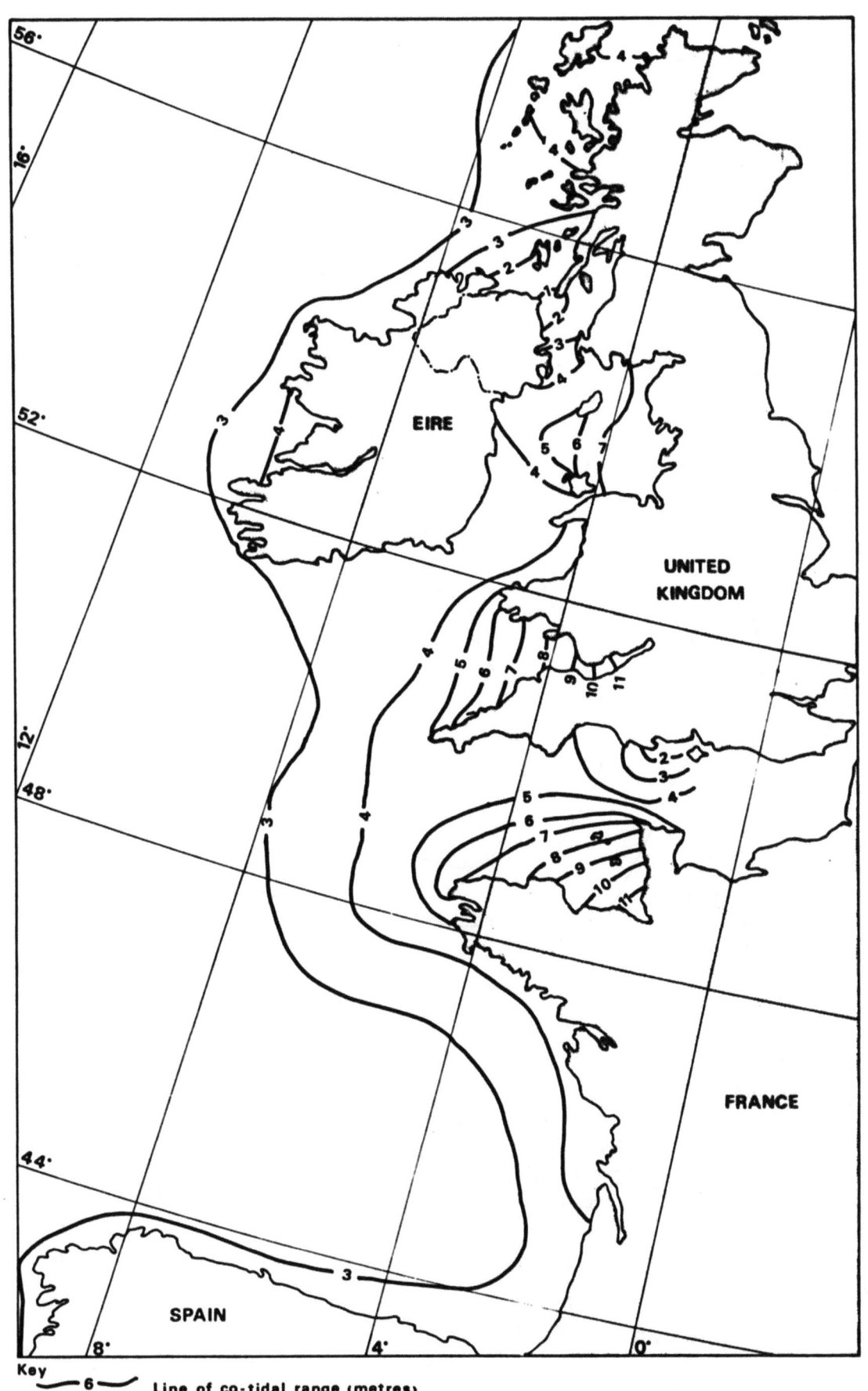

Western approaches-Tidal range Drawing G1

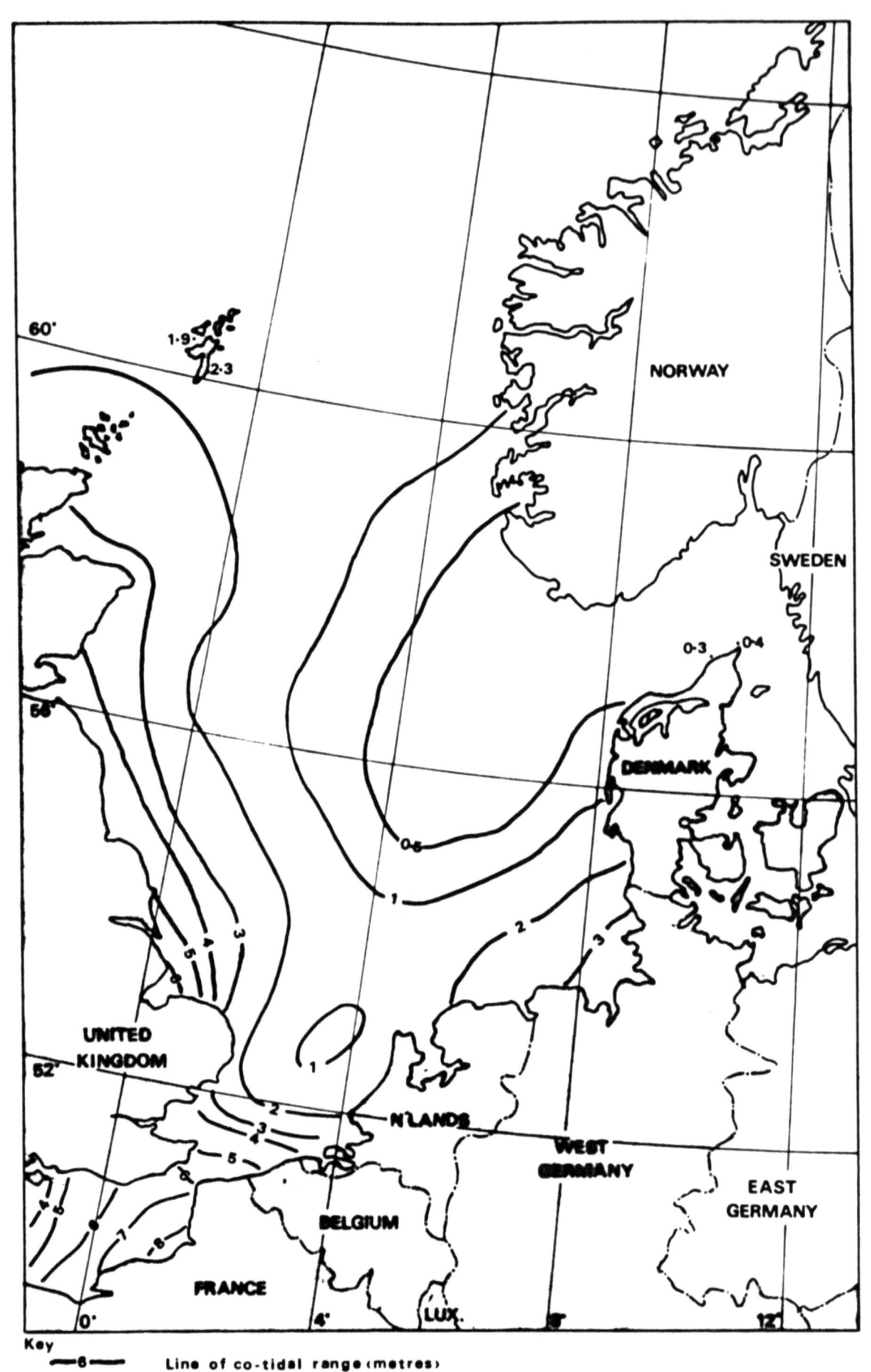

North sea-Tidal range Drawing G2

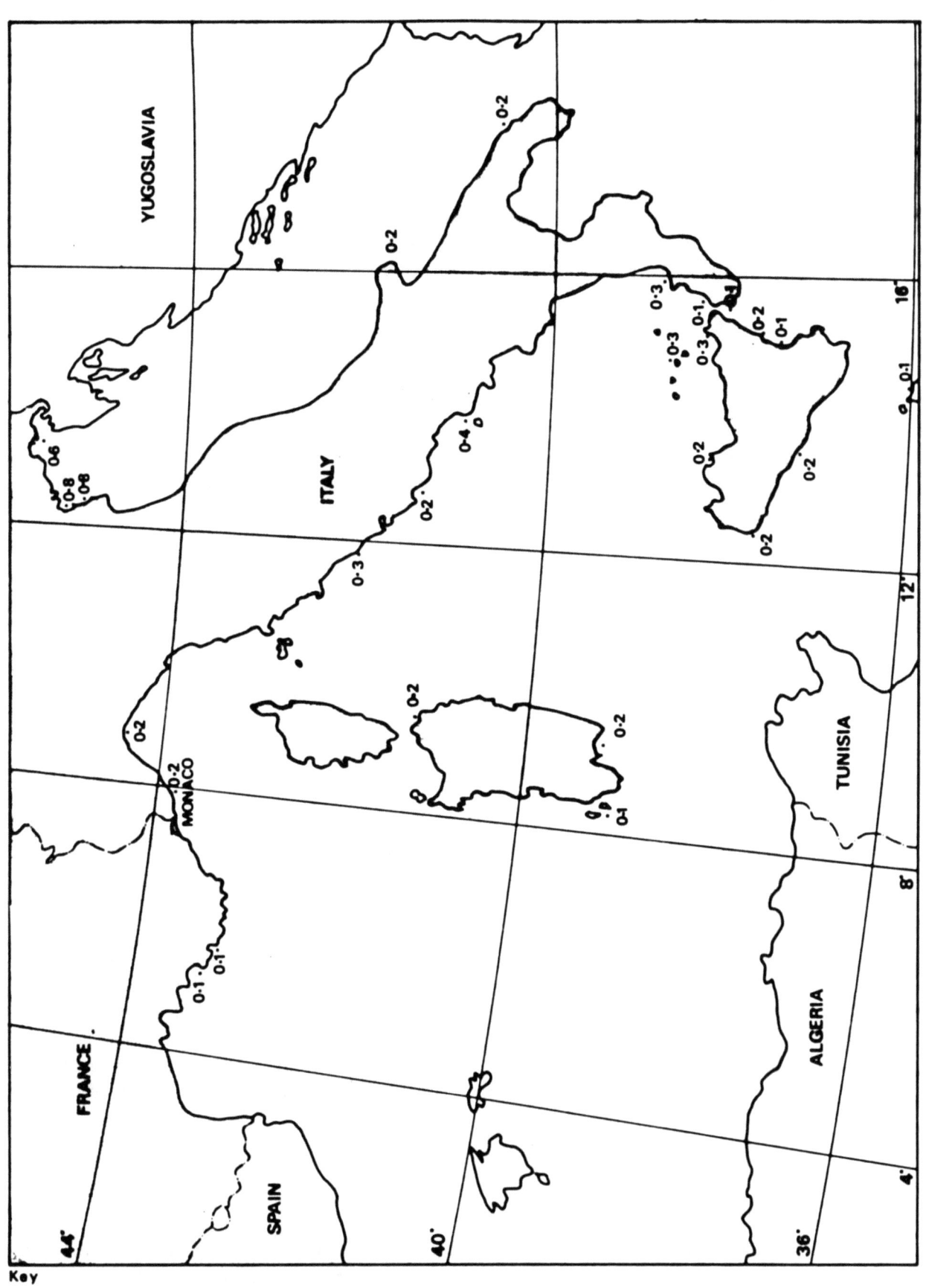

Mediterranean -Tidal range　　Drawing G3

APPENDIX H

EXTREME WATER LEVELS

H1. Introduction

H1.1   In determining both maximum and minimum design water levels for coastal or marine structures, it is necessary to consider the occurrence of extreme levels, beyond the range of predicted tidal range. These occur either as a result of abnormal meteorological conditions, in which case they are known as "storm surges", or as a result of earthquakes disturbing the seabed and causing "Tsunamis'.. The two phenomena are both amplified in shallow water so are of greater significance for fill islands: however they are not related otherwise and have very different characteristics.

H1.2   Since most of Europe is rarely affected by earthquakes, Tsunamis are generally of only minor concern. There would be a need to consider this aspect at design stage but it is unlikely to have much bearing on the choice of site, and only storm surge is considered here in further detail.

H2. Storm surges

H2.1   Surges most commonly of interest are "positive" i.e. those that result in high water levels which are relevant to freeboard allowance. Negative surges are important too, however, as these can affect submergence of intakes, stability of retaining walls and available draft for shipping.

H2.2   There are three principle components of storm surge, viz.

(i) Wind set up

(ii) Increase in level due to reduced atmospheric pressure.

(iii) Wave set up.

The complex dynamic behaviour of storm surge in shallow water makes analysis difficult though surge models are now available to aid prediction of the effect of specified storm events. In other cases, empirical methods are used: in the North Sea, for example, storm surges often follow a recognisable track from North to South and monitoring of their progress, coupled with regression analysis of previous events, provides a good indication of the likely extremes of water level.

H3. Case Histories

H3.1   No analysis is attempted here but table H1 summarises several extreme events which have occurred and which demonstrate the order of water level increase that needs to be considered for offshore islands.

TABLE H1. EXAMPLES OF STORM SURGES AFFECTING EUROPEAN COASTS

Location	Date	Surge residual at H.W.
Dutch coast		
Hook of Holland	1953	3.1 m
German coast		
Cuxhaven	1976	2.3 m
U.K. West Coast		
Morecambe bay	1977	2.1 m
Severn Estuary	1936	2.3 m
U.K. East coast		
Wash Estuary	1978	1.5 m
Thames Estuary	1953	2.1 m
Italy, Adriatic		
Venice	1966	1.9 m

NOTES: 1. Column (3) gives the difference between observed sea level and predicted tide level at time of predicted high water (HW).

2. Surge residuals larger than those given above have been recorded at times other than high water : however these have corresponded to lower absolute levels so were less serious.

APPENDIX J

DYNAMIC STABILITY OF OFFSHORE ISLANDS

J1. Introduction

J.1.1 Many of the forces that act on offshore islands are not steady but fluctuate with time. For example wave forces vary as the water surface rises and falls over a wave cycle; the forces from tidal currents vary in magnitude and direction over the tidal cycle. Other environmental disturbances such as winds and drift currents produce forces which vary irregularly or follow some seasonal pattern.

J2. General principles

J.2.1 The motions and stresses that arise from varying or dynamic loading depend in part on how close the frequency of the disturbing force is to the natural frequency of the structure. For a simple system consisting of a mass M, supported by a spring of stiffness K, the natural frequency $\omega_n = \sqrt{K/M}$. Although an offshore structure is a complex system the basic relations contained in this expression still hold good; the natural frequency increases as the stiffness of the structure increases or its inertia decreases.

J.2.2 When the disturbing force has a frequency close to ω_n for the structure, resonance occurs. In this condition the structure may experience large periodic displacements and stresses. The magnitude of the displacements and stresses at resonance and at other frequencies depends on the magnitude of the disturbing force, the stiffness of the system and the degree to which energy is dissipated and motion damped.

J.2.3 Varying forces which have a very low frequency relative to the natural frequency of the structure may be treated in the same way as steady forces. For offshore structures forces due to currents (excluding localised effects such as vortex shedding) and winds can normally be considered as quasi-steady or drift forces. The predominant dynamic forces on offshore structures arise from wave action. The wave-induced forces are primarily oscillatory and in general have the same frequency bandwidth as the waves themselves.

J3. Floating islands

J.3.1 A floating offshore island has six degrees of freedom. Linear motions along the three co-ordinate axes are called heave, surge and sway. Rotations about these axes are referred to as yaw, roll and pitch respectively. Dynamic forces acting on the island will produce motions in one or more of these modes.

J.3.2 Each mode of oscillation has its own natural frequency. In the absence of mooring forces the natural frequencies of a floating island depend on its mass, the way this mass is distributed and the restoring forces, (e.g. righting moments) that act when the structure is displaced. It is undesirable that the natural frequency of any mode of oscillation should be close to the frequency of waves with a high energy.

Moorings.

J3.3   To prevent the drift forces from moving the island from its initial location some means of maintaining its position is necessary. This could be achieved by a mooring system or by some propulsion that would counteract the drift forces. Unless the island is close to some fixed structure, moderate excursions about its mean position due to drift forces may not be serious.

J3.4   In addition to keeping a floating island within some limited area a mooring system can also modify the dynamic behaviour of an island. By effectively increasing the stiffness of the island the natural frequency of the system can be increased. The amplitude of oscillation and the acceleration of the structure will be reduced by the added stiffness of the moorings. The effect on acceleration may be critical since without mooring forces the accelerations of the structure due to waves may be too high for operation of a nuclear plant.

J3.5   To be effective in limiting the accelerations of a floating nuclear power plant in exposed conditions, the mooring system must be able to exert large forces on the structure. Any initial slack in the system is undesirable since until this slack is taken up the moorings will have no influence on the movement of the structure. It may therefore be necessary to pre-tension the moorings by stressing them against excess buoyancy in the floating structure.

J3.6   As an alternative to or in addition to moorings, the movement and acceleration of a floating structure may be reduced by the provision of breakwaters or other structures (either attached to the island or separate) to absorb wave energy. The configuration of members in the island structure can also be arranged so as to minimise the effect of waves.

J3.7   Detailed consideration of these matters is complex since in practice waves do not arise as regular trains and so the disturbing forces do not follow a simple periodic variation. Simple treatments neglect the non-linear effects that are produced by the waves and damping forces. The extent to which the resistance of the water around the structure modifies its motion is difficult to estimate. The need for the structure to displace an adjacent mass of water is normally allowed for by inclusion of some added masses in the calculations.

J4.   Fixed islands

J4.1   For a fixed offshore structure provided that the dynamic forces do not exceed the ultimate resistance that can be mobilised by the structure no net displacement will occur, at least in the short term. However it is normally desirable to provide a considerable margin of safety to resist cyclical dynamic forces since these may have a cumulative effect. At high enough stress levels fluctuating forces can weaken a structure and lead to its eventual failure. For example, cyclical loading of steel members can lead to fatigue failure or to corrosion fatigue. By ensuring that the fluctuating stresses due to dynamic forces are kept below some lower limiting level, cumulative effects can usually be avoided.

APPENDIX K

COOLING WATER DISPERSION

K1. Introduction

Objectives

K1.1 This brief review outlines the potential problems arising from the discharge of cooling water into the sea, describes the physical processes of heat dispersion and comments on available methods of analysis. A simple method is then used to illustrate the sensitivity of background temperature rise to variations in the important parameters.

Recirculation

K1.2 In designing the cooling water system, it is clearly essential to avoid the situation in which hot water discharged from the outfall is later recirculated via the intake. There are two categories of recirculation: "primary recirculation" arises when hot water from the surface is drawn directly into the intake whereas "secondary recirculation" occurs as a result of an overall temperature rise in the surrounding body of water.

K1.3 In order to avoid recirculation, the principal consideration is the location of intake and outfall. Vertical separation, with the intake drawing in cooler water from a level below the outfall, often provides a method of avoiding primary recirculation though this cannot be guaranteed in shallow water and with large flows. Furthermore the background temperature rise may still be appreciable and horizontal separation may also be required.

K1.4 Provision of an outfall diffuser to distribute the discharge over a length of pipeline increases the initial mixing and hence provides more rapid dilution of the warm water.

Environmental impact

K1.5 A second problem arising from discharge of warm water into the sea is that of thermal pollution. There are many areas of the environment which are sensitive to temperature. The effect of temperature changes on fish, benthic populations, and shoreline fauna would all need to be assessed.

K2. Physical Principles and Methods of Analysis

Physical principles

K2.1 The continuous release of warm water from the outfall will raise the general water temperatures above their natural levels over a large area. The new temperature distribution will depend upon:

- the heat input by the power station.

- the distribution of the cooling water by tidal drift and wind induced currents.

- the mixing of the cooling water with the water in the area in both horizontal and vertical directions.

- the heat losses through the water surface into the atmosphere.

K2.2   It is convenient to consider the mechanism by which the heat is dispersed in two phases (though in practice the processes merge). From the outfall the heated water will spread out over the surface of the sea under the influence of buoyancy, momentum and shear stresses between the two layers. In this "near field" the water will be vertically stratified with the hot water forming a plume, typically about 2 m deep, on the surface of the cooler sea water. Outside a range of one tidal excursion from the outfall, the maximum temperature rise caused by the hot water discharge will be much reduced. The vertical stratification will also diminish progressively until mixing over the water depth is complete and the area then becomes the "far field".

K2.3   Whilst the general picture of the heat dispersion processes are understood, it is very difficult to formulate them in terms of equations which relate parameters that can be defined and measured. In particular the mixing processes, which involve exchange of water at scales varying from molecular diffusion to large ocean eddies, cannot be defined easily. The parameter often used to describe the degree of mixing occurring is known as the dispersion coefficient and is a function of many physical phenomena including density gradients, wave action, tidal currents and seabed topography. It follows that the value of the dispersion coefficient can change significantly in both space and time. Another important variable is the residual drift (ie mean tidal average velocity) and this too can change or even reverse direction according to weather and seasonal variations.

Methods of analysis

K2.4   The complexity of the physical processes which govern the dispersion of heat from cooling water makes it particularly difficult to predict temperature rises resulting for specific cases. Nevertheless, there are methods, of widely varying cost and effort, which can be used to examine the problem. The factor common to all methods is the need for accurate and often extensive field data. Alternative approaches are outlined below and one of these has been selected to demonstrate the sensitivity of temperature change to some of the parameters involved.

K2.5   A number of physical hydraulic models to simulate mixing and surface cooling have been used to investigate specific sites (e.g. Frazer et al, 1968). However, there are many drawbacks to this method, the fundamental problem being that of ensuring hydraulic similarity. Physical models involve high capital cost and are inflexible in that only one site can be studied in each model. The principles and limitations of physical models are described by Ackers (1968).

K2.6   In recent years the use of mathematical models for heat dispersion problems has become increasingly popular and with the development of sophisticated numerical models the method has rapidly become a powerful tool. However, the difficulties of describing the physical processes and the dependence of results on field data impose limits on the reliability of this method too. A particular problem is the knitting together of the "near field" and "far field", which are dealt with using different models.

K2.7   A strength of mathematical modelling is the facility to adapt models to different purposes. Relatively simple models (Ackers et al, 1980) may be used for preliminary studies and to identify potential problems without involving high cost.

K3.   Sensitivity Analysis

Objective

K3.1   Since primary recirculation is dependent not only on marine conditions but also on outfall design and mode of operation, it is virtually impossible to analyse the problem without being site specific. It is however reasonable to believe that, for a given case, it should be possible to avoid the problem at the design stage. The occurrence of secondary recirculation, however, is related almost solely to water depth, current velocities and dispersion coefficients. While these parameters need to be identified for any specific site, it is possible to assess the sensitivity of results to variations in each parameter. This method of analysis has therefore been adopted to indicate the likely range of background temperature increase which would occur near the island for different depths of water. Temperature increases further from the island (for example those that might occur at nearby coastlines) are not assessed here as they have no direct bearing on the outfall design.

Methodology

K3.2   The analysis presented here is based on the analytical approach developed by MacQueen (1978). In the case of intake and outfall being close together (and hence minimising the "mass injection" effect which changes the flow regime) the background temperature rise within the "tidal plug" (an area of length equal to the tidal excursion) for a shoreline outfall is given by

$$\Delta T = \frac{\mu T}{\lambda} \exp(-\lambda y)$$

where: y is the distance offshore

$$\mu = \frac{V}{2a\,h\,Ky}$$

$$\lambda^2 = \left(\frac{E}{\rho_c h} + \frac{Ud}{2a}\right)/Ky$$

where   V is the rate of discharge of cooling water
a is half the tidal excursion
h is water depth
ρ_c is the thermal capacity of water
U_d is residual drift velocity
Ky is lateral dispersion coefficient
E is rate of heat loss to the atmosphere

K3.3 By idealising a mid-ocean outfall as two mirror image shorelines we can obtain a result for a mid-ocean outfall: in this case the design discharge is divided by 2 to give the value of V. Note too that for results applicable near the island, we are only concerned with y = 0.

K3.4 MacQueen's work is also useful in assessing the effect of providing an outfall diffuser to spread the discharge over finite width. In this case the temperature increase, $\Delta T'$, is given by:

$$\Delta T' = \frac{\Delta T}{1+qw}$$

where $q = P - \mu$

$P = \frac{1}{2}\left\{(\mu^2 + 4\lambda^2)^{1/2} + \mu\right\}$

w = diffuser length.

Again the shoreline outfall equation above must be modified for mid-ocean conditions by using

$$w = \frac{\text{diffuser length}}{2}$$

Assumptions and data

K3.5 Discharge of cooling water is taken to be about 50 m³/s per 1000 MW power output. A 2500 MW station is therefore taken to have a discharge of 125 m³/s. The temperature excess is taken as 9K.

K3.6 Rate of heat loss to the atmosphere, E, is taken to be 25W/m² (a value consistent with observed data: This value could vary by a factor of two or more according to prevailing weather conditions but is taken as constant here).

K3.7 The values of residual drift velocity, U_d, and dispersion coefficient, Ky, are not only dependent on location but may also vary seasonally. Measurement of both parameters is, in any case, time consuming and expensive, so it is only possible to present a range of data likely to be encountered and hence to demonstrate the sensitivity of ΔT to these variations.

K3.8   Tidal excursion, 2a, can be measured easily at any site but again this varies with location and throughout the spring-neap tidal cycle.

K3.9   Table K1 summarises the ranges of values assumed for the principal parameters. Two types of site have been considered: one with conditions characteristic of the North Sea or the Western Approaches, the other with conditions likely to be found in the Mediterranean.

K3.10   Table K2 shows the constant used for the sensitivity calculations.

	North Sea and Western Approaches	Mediterranean
Tidal excursion (2a)	10 - 30 km	2 - 10 km
Dispersion coefficient (Ky)	1.0 - 5.0 m²/s	0.2 - 2.0 m²/s
Residual drift velocity	0.02 - 0.20 m/s	0.02 - 0.20 m/s

Table K1: Ranges of values for principal parameters

Discharge rate	125 m³/s
Temperature	9K above ambient sea water
Heat loss to atmosphere	25 W/m³/K

Table K2: Basic parameter values — constant for this study

K3.11   Results are plotted as graphs of ΔT as a function of drift velocity, U_d, for water depths of 5, 20 and 50 m. Figures 1a-1d are for results representing a range of tidal excursions and dispersion coefficients typical of UK Coastal waters in the North Sea. It is also reasonable to consider these as broadly representative of other relatively exposed coasts designated Western Approaches. Conditions in Mediterranean waters are very different however and data is less easily available. In the absence of data, realistic values can only be estimated roughly but figures 2a–2d indicate the possible range of ΔT for various combinations of tidal excursion and dispersion coefficient.

K3.12   The effect of an outfall diffuser is demonstrated in figure 3 which shows, for fixed values of depth, dispersion coefficient and tidal excursion, the temperature rise for three different diffuser lengths.

K4.   Conclusions

K4.1   Careful design and siting of the intake and outfall for the cooling water system are required in order to avoid problems of recirculation and of excessive temperature rises which might affect environmentally sensitive areas. The risk of primary recirculation can generally be minimised by vertical separation of intake and outfall (i.e. by utilising the stratification of hot and cold water which results from the density difference between the two): it is assumed, therefore, that secondary recirculation is more likely to be of concern, and it is only this aspect that is considered here.

K4.2   The likely extent and duration of background temperature rises resulting from discharge of cooling water depends on many factors of which water depth, tidal currents and dispersion coefficients are the most important. The relevant parameters vary significantly from one area to another and it is only possible to predict results for a specific site when extensive data have been acquired.

K4.3   A simple analytical method has been adopted to demonstrate the sensitivity of results to changes in the principal parameters. The results of the analysis are presented graphically and show that:

(i)   The small tides and correspondingly low tidal excursion and mixing in Mediterranean waters create less satisfactory conditions for dispersion than in the North Sea or Western Approaches.

(ii) Sites in shallow water are particularly susceptible to high background temperature rises.

(iii) Provision of an outfall diffuser is effective in reducing background temperature rises.

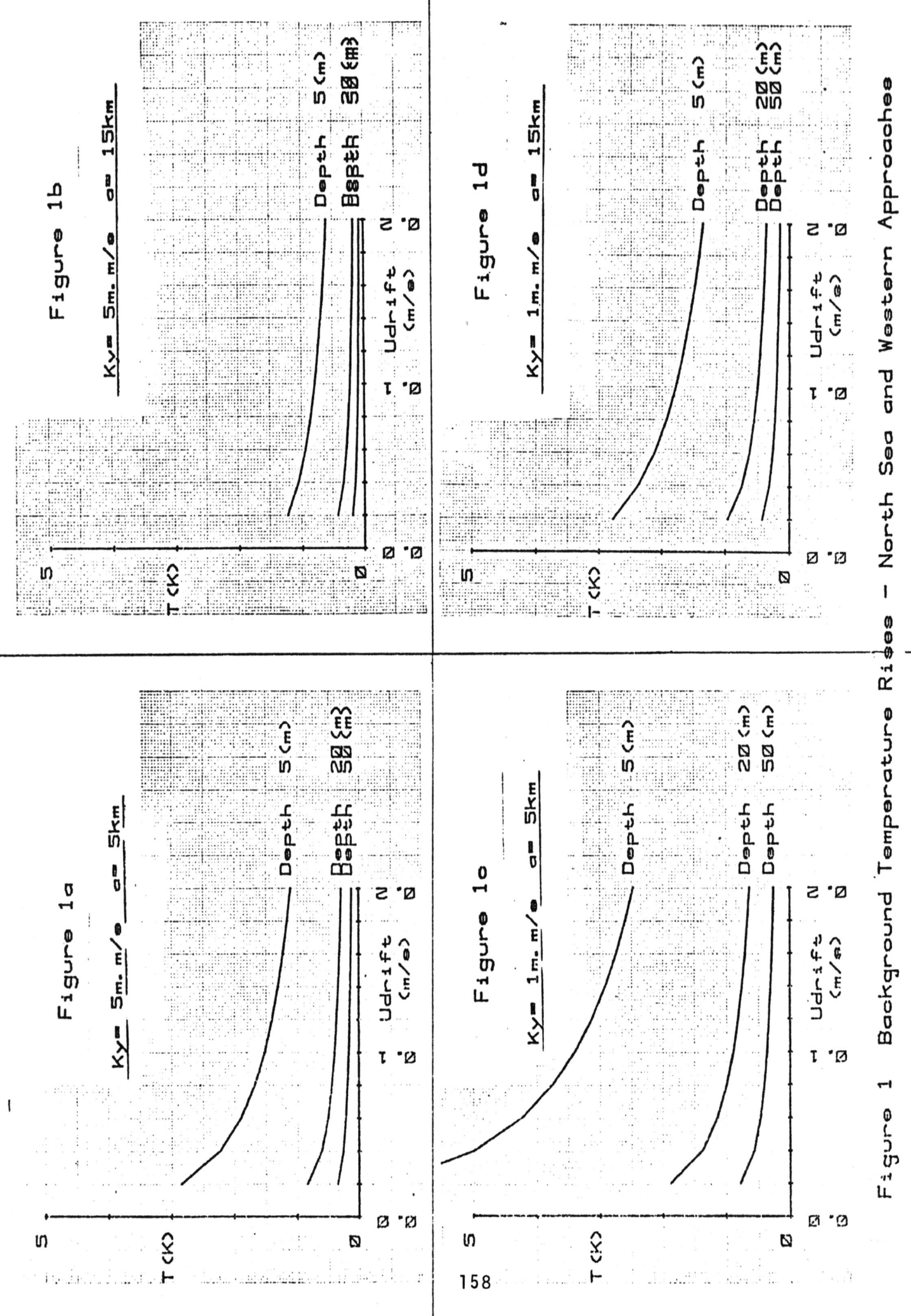

Figure 1 Background Temperature Rises - North Sea and Western Approaches

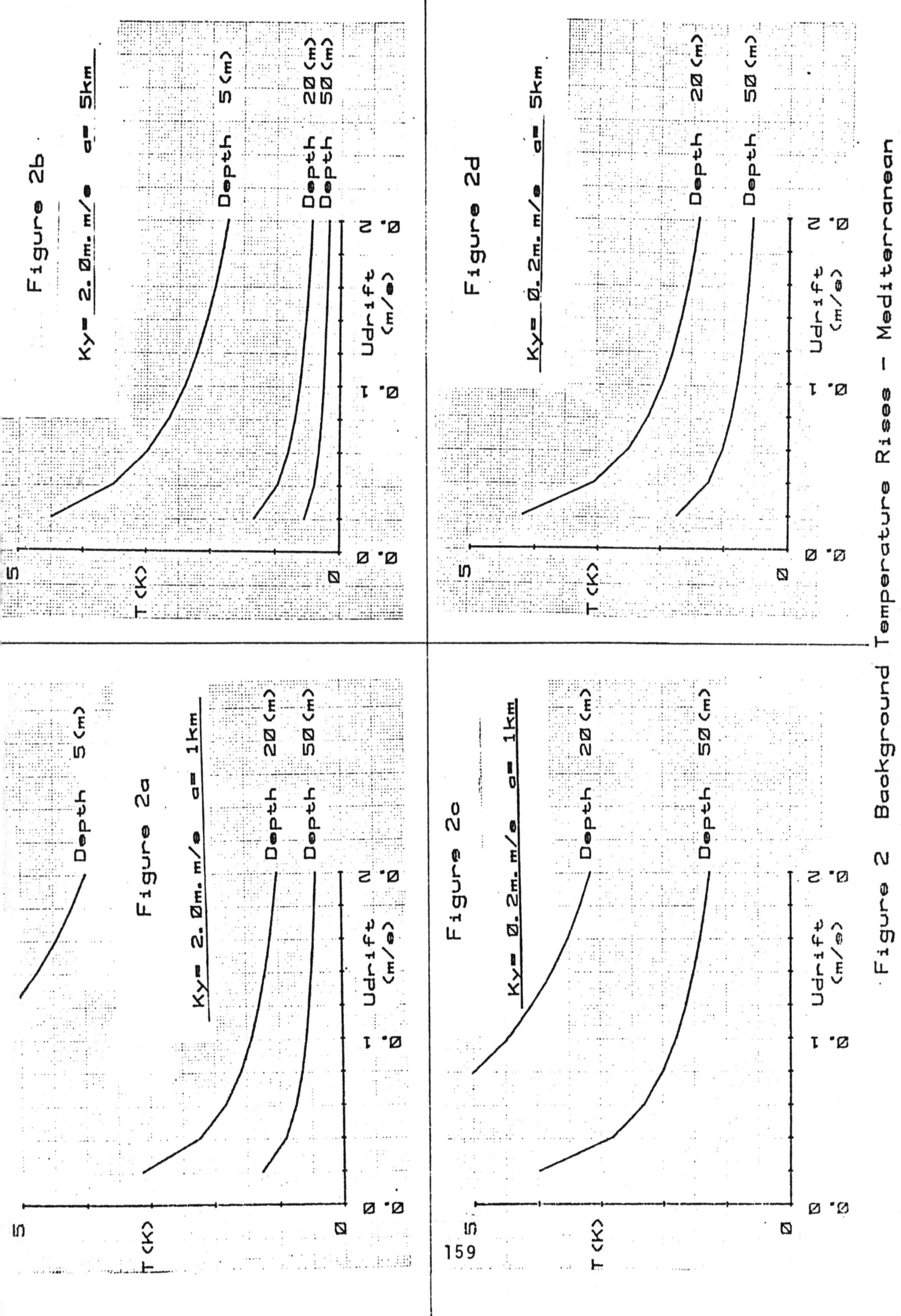

Figure 2 Background Temperature Rises - Mediterranean

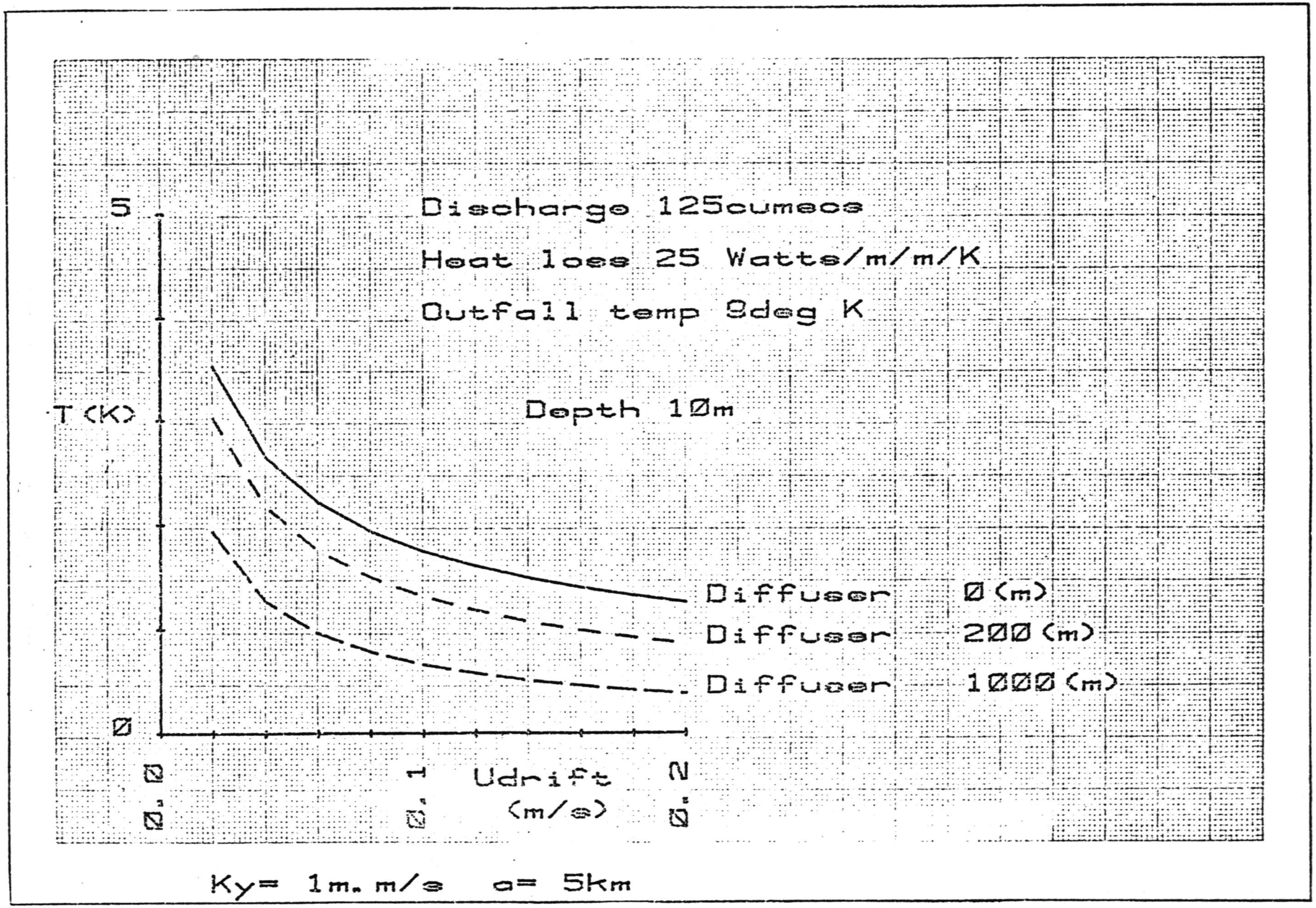

Figure 3 Effect of outfall diffuser

If you have any concerns about our products,
you can contact us on
ProductSafety@springernature.com

In case Publisher is established outside the EU,
the EU authorized representative is:
**Springer Nature Customer Service Center GmbH
Europaplatz 3, 69115 Heidelberg, Germany**

Printed by Libri Plureos GmbH
in Hamburg, Germany